StatSheets with 3.5″ Excel Templates and Data Files

for use with

Complete Business Statistics

Fourth Edition

Amir D. Aczel
Bentley College

Prepared by
Jayavel Sounderpandian
University of Wisconsin — Parkside

Irwin McGraw-Hill

Boston Burr Ridge, IL Dubuque, IA Madison, WI New York San Francisco St. Louis Bangkok Bogotá Caracas Lisbon London Madrid Mexico City Milan New Delhi Seoul Singapore Sydney Taipei Toronto

Irwin/McGraw-Hill

A Division of The McGraw-Hill Companies

StatSheets with 3.5″ Excel Templates and Data Files for use with
COMPLETE BUSINESS STATISTICS

2 3 4 5 6 7 8 9 0 BKM/BKM 9 3 2 1 0 9

ISBN 0-07-289306-0

http://www.mhhe.com

Preface

Prerequisites

The spreadsheet templates presented here assume that the reader will use Microsoft Excel 97 or later version running on a Pentium or better computer. It is also presumed that the reader has the basic knowledge of spreadsheets, such as entering or copying formulas and data.

The Issue of Black Box

A danger of using spreadsheet templates in education is that the students may use them mechanically without understanding the concepts behind the techniques. To minimize this danger,

<u>If you are a student</u>:

- Do not fail to learn the concepts.
- While working with templates, click on important cells to see the formulas in them. Make sure you understand those formulas.
- Experiment with the data input to see how the answer changes when data changes.
- Try to solve a problem without using a template but using a blank spreadsheet.
- See if you can construct your own templates.

<u>If you are an instructor</u>:

- Do not fail to teach the concepts behind every template.
- Do not fail to test the students for their understanding of concepts. An exam should contain many conceptual questions as well as problems.
- Problems should address how the answer would change if the data is changed.
- Make the students solve at least some problems without the use of templates.
- Assign projects that involve development of new or modified templates.

Disclaimer

Although I have worked with all of the templates presented here and checked them for errors, some errors may still remain in any of them. You should note that you are using the templates at your own risk. Neither I nor the publishers assume any responsibility for consequences of errors in these templates.

I would, of course, like to know any error that you spot in the templates or in the write up. Please contact me (Sounderpandian@hotmail.com) about the errors you find.

JAYAVEL SOUNDERPANDIAN
University of Wisconsin-Parkside
July 1998

IRWIN/McGRAW-HILL

Statsheets with 3.5" Excel Templates and Data Files for use with Complete Business Statistics, 4/e, by Amir D. Aczel.

Please use this postage-paid form to report any errors that you find in this material. Be as complete as possible noting specifically which changes should be made. We will address them in subsequent printings and future editions. Thank You.

NOTE: Extra copies of this form appear at the end of this manual.

Attention: Scott Isenberg

Name ____________________________ School ____________________________

Office Phone ____________________________

Please fold and seal so that our address is visible.

NO POSTAGE
NECESSARY
IF MAILED
IN THE
UNITED STATES

BUSINESS REPLY MAIL
FIRST-CLASS MAIL PERMIT NO. 204 OAKBROOK, IL

POSTAGE WILL BE PAID BY ADDRESSEE

ATTENTION: **Scott Isenberg**

THE McGRAW-HILL COMPANIES
RICHARD D. IRWIN
1333 BURR RIDGE PKY.
BURR RIDGE, IL 60521-0085

(fold)

(fold)

Contents of the Diskette Enclosed

The diskette enclosed with this booklet contains all the spreadsheet templates discussed in the chapters. This table presents the contents of each of the workbooks, in the order in which they are presented in the chapters.

Workbook	Sheet	Function/Contents
Descriptive Statistics	Basic Stats	Calculates basic statistics such as percentiles, mean, variance from raw data.
	Box Plot	Creates a box plot from raw data.
	Project 1	To be used for a project on basic statistics.
	Project 2	To be used for a project on moments.
	Project 3	To be used for a project on SSD.
Histogram	Raw Data	Creates a histogram from raw data.
	Grouped Data	Creates a histogram from grouped data.
Probability	SuccessProbs	Calculates the probability of at least one success in many independent trials.
	Joint Prob Table	Creates a Joint Probability Table from Crosstabulation.
Distributions	Discrete	Describes a discrete probability distribution.
	E[h(x)]	Calculates E[h(x)] where x is a random variable.
	Descriptive Stats	Calculates descriptive statistics of a random variable.
	Binomial	Calculates binomial probabilities.
	Negative Binomial	Calculates negative binomial probabilities.
	Geometric	Calculates geometric probabilities
	Hypergeo	Calculates hypergeometric probabilities.
	Poisson	Calculates Poisson probabilities
	Uniform	Calculates uniform distribution probabilities.
	Exponential	Calculates exponential probabilities.
Normal Distribution	Normal	Normal distribution calculations.
	Normal approx. of Binomial	Calculations based on normal approximation of binomial probabilities.
Confidence Intervals	CI for mu	Confidence Intervals for population mean.
	t Distribution	t Distribution calculations.
	CI for p	Confidence Intervals for population proportion.
	Chi Square Distn.	Chi Square distribution calculations.
	CI for Popn. Variance	Calculates confidence intervals for population variance.
	Sample size	Calculates minimum sample sizes for estimating population means and proportions.
Hypothesis Testing	Tests of mu	Testing hypotheses about population means.
	z-Test for p	Testing hypotheses about population proportions.
	Binomial test for p	Testing hypotheses about population propotions using binomial probability calculation. This test is more accurate than the *z*-test.
	Test for Popn. Var.	Testing hypotheses about population variance.
	Power of a mu test	Calculates the power of a hypothesis test about population mean.
	Power of a p test	Calculates the power of a hypothesis test about population proportion.
	n for Testing mu	Sample size determination for testing hypotheses about population means, controlling Type I and Type II error probabilities.

Hypothesis Testing (Continued)	n for Testing p	Sample size determination for testing hypotheses about population proportions, controlling Type I and Type II error probabilities.
	n for Testing p Binomial	Sample size determination for testing hypotheses about population proportions, controlling Type I and Type II error probabilities calculated using binomial distribution. More accurate than normal distribution calculations.
Testing Differences	z-Test	Testing differences between two population means
	t-Test Equal Variance	Testing differences between two population proportions assuming equal population variances.
	t-Test Unequal Variance	Testing differences between two population proportions assuming unequal population variances.
	t-Test Paired Difference	Paired difference test for population means.
	Popn. Proportions	Testing differences in two population proportions.
	F Distn.	F distribution calculations.
	F-Test	Testing hypotheses about equality of two population variances.
ANOVA	Concepts	Explains 1-way ANOVA concepts.
	DEVSQ	Explains the use of DEVSQ function for ANOVA.
	Unequal	1-way ANOVA with unequal number of observations.
	ToolPak	Use of Analysis ToolPak for 1-way ANOVA.
	ToolPak2	Use of Analysis ToolPak for 1-way ANOVA with unequal number of observations.
	2 Factor	2 Factor ANOVA calculations.
	2 Factor ToolPak	Use of Analysis ToolPak for 2 Factor ANOVA
	Rand. Comp. Block	Randomized Complete Block Design ANOVA
	RCB ToolPak	Use of Analysis ToolPak for Randomized Complete Block Design ANOVA.
Simple Regression	Scatter Plot	Constructs Scatter Plot from raw data.
	Visual	Visually solving simple regression problem.
	Solver	Use of the Solver for simple regression.
	LINEST	Use of LINEST function for simple regression.
	Matrix	The matrix method for simple regression.
	Analysis ToolPak	Use of Analysis ToolPak for simple regression.
	Normal Eqns.	The normal equations method for simple regression.
	Trendline	The Trendline method for simple regression.
	Formulas	Excel formulas for simple regresson.
	Norm. Prob. Plot	Creates a normal probability plot of residuals.
Multiple Regression	LINEST	Use of LINEST function for multiple regression.
	Solver	Use of the Solver for multiple regression.
	Data	Enter the data for a multiple regression here and …
	Results	read the results here, and …
	Residuals	look at the residual plot here.
Forecasting	Trend	Explains trend calculation.
	Moving Avg.	Calculates moving averages.
	Exp. Smoothing	Exponential smoothing calculations.
	Trend+Season	Forecasting with Trend and Seasonality in data.
	Index	Index calculations.
Quality Control	Pareto	Creates a Pareto chart from frequency data.
	x-bar R s charts	Creates Mean, Range and Stdev control charts from raw data.
	p Chart	Creates p control chart from raw data.
	c Chart	Creates c control chart from raw data.

Nonparametric Tests	Sign	Sign test from raw data.
	McNemar	McNemar test.
	Mann-Whitney	Conducts Mann-Whitney test from raw data.
	Wicoxon	Conducts Wilcoxon Signed Rank test from raw data.
	Kruskal-Wallis	Conducts Kruskal-Wallis test from raw data.
	Friedman	Conducts Friedman test from raw data.
	Spearman	Conducts Spearman Rank Correlation test from raw data.
	Chi-Sq. for Fit	Conducts a chi-square test for goodness-of-fit.
	Chi-Sq. for Ind.	Conducts a chi-square test for indepence.
Decision Analysis	Without Probs	Decision analysis without probabilities.
	With Probs & Info	Decision analysis with probabilities and additional information.
	Marbles	Explains Bayes' Rule.
	Data	Enter the data for a decision analysis problem here and ..
	Results	Read the results here.
	Calc	Contains the decision analysis calculations.
Sampling Methods	Stratified for mu	Estimating population mean by Stratified sampling.
	Startified for p	Estimating population proportion by Stratified sampling.
	Allocation	Optimal allocation for stratified sampling
	Cluster for mu	Estimating population mean by Cluster sampling.
	Cluster for P	Estimating population proportion by Cluster sampling.
	Systematic	Estimating population mean by Systematic sampling.

TABLE OF CONTENTS

0. Some Advanced Features of Excel

This chapter contains descriptions of some advanced features of Microsoft Excel (Excel 97 or later) that are used in this book.

0.1 CTRL+Enter and CTRL+SHFT+Enter

Suppose you need to enter the same or similar formula into many cells. The copy command along with absolute references using $ signs as necessary can be used to carry out such tasks. An easier method is the use of `CTRL+Enter` and `CTRL+SHFT+Enter` keystrokes. [`CTRL+Enter` means that the `CTRL` key is kept pressed and the `Enter` key is punched, and `CTRL+SHFT+Enter` means that both `CTRL` and `SHFT` keys are kept pressed and the `Enter` key is punched.]

Let us say we want to create the sheet in Figure 0.1.1 where column B displays values of X, column C twice the values of X, and column D twice the value in cell B3.

	A	B	C	D
1				
2		X	2X	2*B3
3		10	20	20
4		20	40	20
5		30	60	20
6		40	80	20

Figure 0.1.1 CTRL+Enter and CTRL+SHFT+Enter

One way to create column C is to enter the formula `=2*B3` in cell C3 and copy it into C4:C6. A faster way is:

- Select the range C3:C6
- Enter the formula `=2*B3`
- Press `CTRL+Enter`

This will have the same effect as before.

As for column D, one way to do it is to enter the formula `=2*$B$3` in cell D3 and copy it into D4:D6. A faster way is to use the *array formula* using `CTRL+SHFT+Enter` feature.

- Select the range D3:D6
- Enter the formula `=2*B3` (Note: no dollar sign is needed)
- Press `CTRL+SHFT+Enter`

Notice how curly brackets { } appear around the formula when you use the `CTRL+SHFT+Enter` feature. Formulas entered in this manner are called *array formulas.* The cells into which the formula is entered is becomes an *array*, which is treated as a single unit. You cannot change or delete the formula in part of an array; if you try, the message "Cannot change part of an array" would appear and nothing else would happen.

An array can be any rectangular range of cells. An array formula can produce different values in different cells of the array. It can be difficult at times to know where the array starts and ends. If you are on one of the cells of an array and want to see the whole array, press `CTRL+/` and the whole array would be selected. In the above sheet, clicking on D4 and pressing `CTRL+/` will select the range D3:D6 revealing that the array consists of all the cells in the range D3:D6.

0.2 Naming Cells or Ranges

0.2.1 Naming Cells

The name of the active cell appears in a box on the left end of the Formula Bar, known as the *Reference Area* or *Name box.* In Figure 0.2.1, the active cell is B3, and its name B3 appears in the reference area. For convenience, one might want to assign another name, a descriptive one, to this cell and refer to it by that name in formulas. For example, cell B3 can be assigned the name "Profit". To do it,

- Select cell B3 by clicking on it.

- Click on the reference area.
- Type `Profit` and press `Enter`.

Now, the formula `=2*Profit` in cell B4 yields twice the profit, which is twice cell B3. The formula `=2*Profit` is more comprehensible than `=2*B3` and that helps in understanding the calculation or debugging.

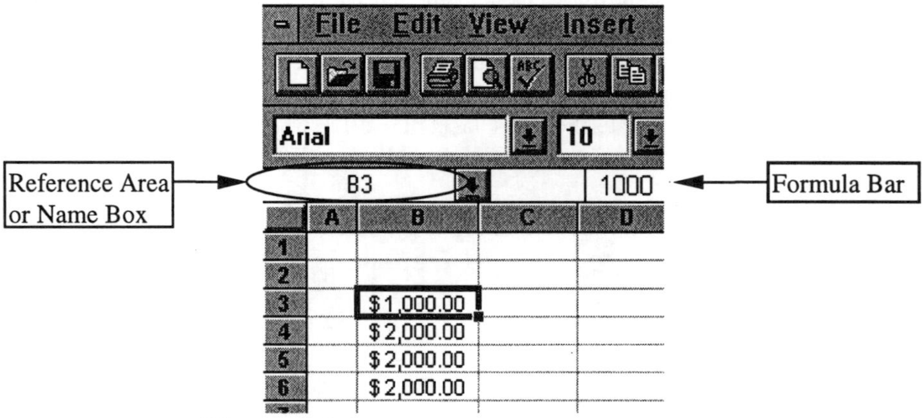

Figure 0.2.1 Formula Bar and Reference Area

Another advantage of names stems from the fact that when a formula containing a name is copied, the name stays as it is. In Figure 0.2.1, when the formula in cell B4 is copied into cells B5 and B6, the formula stays as `=2*Profit` yielding $2,000.00 in each of those cells. Thus the need for absolute reference using $ signs is avoided.

A more important advantage of names is demonstrated in the next example. In Figure 0.2.2, the average, the sum, the maximum and the minimum of the data in the range A2:A7 have been calculated. Suppose we do not name any cells. The formula in D2 will then be `=AVERAGE(A2:A7)` and similar formulas will be in cells D3 to D5. Now let us say we receive an extra data point and add it at cell A8. Because of this addition we need to revise the calculations. The revision would require us to modify the formulas in each of the cells D2 to D5. That is, the formula in D2 will have to be changed to `=AVERAGE(A2:A8)` by changing A7 to A8, and similar changes will have to be made in each of the cells D3, D4 and D5. This task is clearly tedious and warrants some simplification.

Suppose we had named the range of cells (A2:A7) as `Data` and had done the calculations by entering the formula `=AVERAGE(Data)` in D2 and similarly in cells D3, D4 and D5. Now the revision would only require redefining the name `Data` to refer to the range A2:A8. This can be done as follows.

- Under the **Insert** menu, select the command **Name**, and **Define**.
- In the dialog box that appears, click on the name `Data`.
- In the **Refers To** box at the bottom, enter `A2:A8` and click the **OK** button.

Since `Data` now refers to A2:A8, the formulas in cells D2 to D5 are automatically correct, without any revision! Indeed, the name `Data` could have been used in any number of places and all of them would work correctly. Naming cells can thus be quite useful.

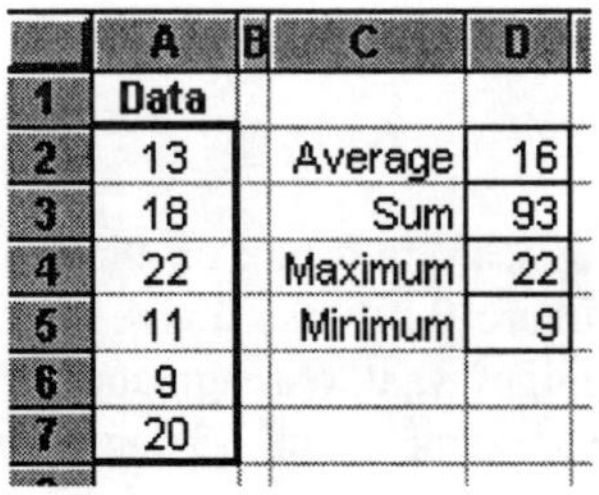

Figure 0.2.2 Use of Names

0.2.2 Combining Names and Array Formulas

In Figure 0.2.3, we have *x* and *y* values given. The *x* - *y* values have been calculated using the formula `=A4-B4` in cell C4 and copying it into the range C5:C9. The sum of all *x* - *y* values are calculated in cell C10 using the SUM formula. This way of calculating the sum of all *x* - *y* values is not only tedious but can also be annoying if we have to add one more data point under *x* and *y*. The task can be much simplified by naming the *x* and *y* values in the ranges A4:A9 and B4:B9 as `x` and `y` and using the *SUM* array formula `=SUM(x-y)` entered by pressing `CTRL+SHFT+Enter`. This has been done in cell C12. This not only avoids the need for the *x* - *y* column, but also makes addition of further data points easy. When the names, `x` and `y` are redefined to include additional data points, the SUM array formula will return the correct answer.

	A	B	C	D
1	SUM array formula			
2				
3	x	y	x - y	
4	12	10	2	=A4-B4
5	14	9	5	
6	15	3	12	
7	13	6	7	
8	17	8	9	
9	18	11	7	
10			42	=SUM(C4:C9)
11				
12			42	{=SUM(x-y)}

Figure 0.2.3 Use of SUM Array Formula

More complicated SUM array formulas are possible with IF and SUMIF functions, and they will be used in later chapters.

0.3 The Data/Table Command

Often a situation calls for looking at many alternatives at once for easy comparison. A table created in Excel will be very helpful in these instances.

In Figure 0.3.1, sales figures have been calculated for years 2001 to 2005 using an annual growth rate of 2%, starting with a sales of 316 in year 2001. [In cell C5, the formula `=B5*(1+$C$2)` has been entered and copied to the right.] Suppose we are not sure about the growth rate and believe it may be anywhere between 2% and 7%. Suppose further, we are interested in knowing what the sales figures would be in the years 2004 and 2005 at different growth rates. In other words, we want to input many different values in cell C2 and see its effect on cells E5 and F5. The table in the range D8:F13 summarizes just that.

	A	B	C	D	E	F
1						
2	Annual growth rate		2%			
3						
4	Year	2001	2002	2003	2004	2005
5	Sales	316	322	329	335	342
6						
7				Growth Rate	2004 Sales	2005 Sales
8				2%	335	342
9				3%	345	356
10				4%	355	370
11				5%	366	384
12				6%	376	399
13				7%	387	414

Figure 0.3.1 Creating a Table

To create this table,

- Enter the growth rates `2%, 3%` etc. in the range D8:D13.
- Enter the formula `=E5` in cell E8 and `=F5` in cell F8.
- Select the range D8:F13.
- Under the **Data** menu select the **Table** command.
- In the dialog box that appears, in the **Column Input Cell** box, type `C2` and press `Enter`. [We use the **Column Input Cell** rather than **Row Input Cell** because the input values are in a column, in the range D8:D13.]

The desired table now appears in the range D8:F13. It is worth noting here that this table is "live" meaning that if any of the input values in the range D8:D13 is changed the table is immediately updated. Also, the input values (in the range D8:D13) could have been calculated using formulas, or could have been a part of another table in the left.

In general, the effect of changing the value in *one* cell, called the *changing cell,* on the values of one or more other cells, called the *result cells,* can be tabulated using the **Table** command.

At times, we may want to tabulate the effect of changing *two* cells. In this case we can tabulate the effect on only *one* result cell, using a 2-Dimensional table. Suppose in the previous example, in addition to the growth rate we are also not sure about the starting sales figure of 316 in year 2001, and believe it may be anywhere between 316 and 324. Suppose further that we are interested only in the sales figure for year 2005. A table varying both the growth rate and the year 2001 sales has been calculated in Figure 0.3.2.

To create the table in Figure 0.3.2,

- Enter the input values for growth rate in the range B9:B14.
- Enter the input values for year 2001 sales in the range C8:G8.
- Enter the formula `=F5` in cell B8. [This is because we are tabulating what happens to cell F5.]
- Select the range B8:G14.
- Under the **Data** menu select the **Table** command.
- In the dialog box that appears enter `B5` in the **Row Input Cell** box and `C2` in the **Column Input Cell** box and press `Enter`.

The table appears in the range B8:G14. This table is also "live" in that any of the input values can be changed and the table updates immediately.

	A	B	C	D	E	F	G
1							
2	Annual growth rate		2%				
3							
4	Year	2001	2002	2003	2004	2005	
5	Sales	316	322	329	335	342	
6							
7					2001 Sales		
8		342	316	318	320	322	324
9		2%	342	344	346	349	351
10		3%	356	358	360	362	365
11	Growth	4%	370	372	374	377	379
12	Rate	5%	384	387	389	391	394
13		6%	399	401	404	407	409
14		7%	414	417	419	422	425

Figure 0.3.2 Creating a 2-Dimensional Table

The appearance of 342 in cell B8 is distracting. It can be hidden by either changing the text color to white or formatting the cell with `;;`. Suitable borders also improve the appearance of the table.

0.4 Goal Seek Command

The **Goal Seek** command can automate a trial and error process of making the value in a particular *target cell* reach a "goal" by changing the value in another *changing cell.* Clearly, the value in the target cell must depend on the value in the changing cell for this scheme to work. In the previous example, suppose we are interested in finding the growth rate that would attain the goal of a sales figure of

400 in the year 2005. [We assume 2001 sales to be 316.] One way to do it is to manually change the growth rate in cell C2 up or down until we see 400 in cell F5. But this task is tedious. So we automate it using the **Goal Seek** command as follows.

- Under the **Tools** menu select the **Goal Seek** command. A dialog box appears.
- In the **Set Cell** box enter `F5`.
- In the **To Value** box enter `400`.
- In the **By Changing Cell** box enter `C2`.
- Click on the **OK** button.

The computer makes thousands of trials and stops when the value in cell F5 equals 400 accurate to several decimal places. The value in cell C2 is the desired growth rate, and it shows up as 6.07%.

0.5 The Solver Tool

The **Solver** tool is a giant leap forward from the **Goal Seek** command. It can be used to make the value of a target cell equal a predetermined value or, more commonly, reach its maximum or minimum possible value, by changing the values in many other cells. In addition, many constraints can be imposed on the values of selected cells, such as restricting a changing cell's value to be between, say, 10 and 20. Let us see the use of the tool through an example.

A production manager finds that the cost of manufacturing 100 units of a product in batches of x units at a time is $4x + 100/x$. She wants to find the most economic batch size. [This is a typical economic batch size problem, famous in production planning.] In solving this problem, we note that there is a constraint, namely, x must be between 0 and 100. Mathematically, we can express the problem as:

Minimize $4x + 100/x$
Subject to $x >= 0$
$x <= 100.$

We set up the problem as shown in Figure 0.5.1.

	A	B	C	D
1				
2		Batch size	10	
3		Cost	50	=4*C2+100/C2

Figure 0.5.1 Solver Application

In cell C3 the formula `=4*C2+100/C2` has been entered. The batch size in cell C2 can be manually changed and the corresponding cost can be read off from cell C3. A batch size of 20, for instance, yields a cost of 85; a batch size of 2 yields a cost of 58. To find the batch quantity that has the least cost, we can use the **Solver** as follows.

- Under the **Tools** menu select the **Solver** command.
- In the **Solver** dialog box, enter `C3` in the **Set Cell** box
- Click on the minimum button, because we want to minimize the cost.
- Enter `C2` in the **Changing Cells** box.
- Click on the **Add** button to add a constraint.
- In the dialog box that pops up, click on the left-hand-side box and enter `C2`.
- In the middle drop down box, select **>=**.
- Click on the right-hand-side box, enter `0` and click on the **ADD** button. [We click on the **ADD** button because we have one more constraint to add.]
- In the new dialog box that appears, click on the left-hand-side box and enter `C2`.
- In the middle drop down box, select **<=**.
- Click on the right-hand-side box, enter `100` and click on the **OK** button. [The **Solver** dialog box should re-appear and it should look like the Figure 0.5.2.]
- Click on the **Solve** button.

The Solver carries out a sophisticated computation process internally and finds the solution, if one exists. When the solution is found, it asks if you want to keep the solution or retain the original values for

When the data set is of a sample, we do not know the true population mean μ. We are forced to calculate the deviations from the sample mean $\bar{x}$ rather than the true mean μ, and this introduces a small downward bias in the sum of all the squared deviations, SSD. To compensate for this bias, we divide SSD by $(n-1)$ rather than n to find the average. The result is the *sample variance* and it is denoted by s^2. We can write the formula for sample variance as

$$s^2 = \text{SSD}/(n-1)$$

Manual computation of variance is tedious since each deviation and its square have to be calculated, and then averaged. On the spreadsheet the functions VAR and VARP can be used to calculate the sample and population variances respectively. In Figure 1.1.1, cell H11 has the formula

`=VAR(Data)`

and calculates the sample variance as 19.9237; cell I11 has the formula

`=VARP(Data)`

and calculates the population variance as 18.9275.

There is a small problem in using variance as the measure of variability. The deviations are squared during the calculation, and therefore, if we started with our data in, say, 'dollars', then the variance would be in 'dollars squared'. In a subsequent analysis, 'dollar squared' may not be appropriate and we may want it to be 'dollars'. To get back to the original unit of dollars, we take the (positive) square root of the variance and call it the *standard deviation.* As can be guessed, the population standard deviation is denoted by σ and the sample standard deviation is denoted by s.

Definition 1.3.5: The *standard deviation* of a data set is the (positive) square root of the variance.

Excel has the formulas STDEV and STDEVP to calculate sample and population standard deviations respectively. In Figure 1.1.1, cell H12 has the formula

`=STDEV(Data)`

and calculates the sample standard deviation as 4.4636. Cell I12 has the formula

`=STDEVP(Data)`

and calculates the population standard deviation as 4.3506. Note that these two standard deviations could have also been calculated by simply taking the square root of the two variances. In other words, cell H12 could have the formula `=SQRT(H11)` and it could be copied into I12.

1.3.1 The *z*-Score

Because the standard deviation is in the same units as the original data, it can be used to *standardize* the deviation of any data point from the mean. Suppose the mean of a population data set is 1000 and the standard deviation is 10. A data point in the set is, say, 1012. The deviation of this data point from the mean is 12. We divide this 12 by the standard deviation of 10 to get 1.2. In other words, the data point is 1.2 standard deviations more than the mean. Then the *z-score* of the data point is said to be 1.2. Similarly, a data point of 986 has a z-score of -1.4 = (986-1000)/10. The formula for z-score is

$$z = (x - \mu)/\sigma$$

where x is the data point, μ is the mean and σ is the standard deviation.

On a spreadsheet, the z-score is calculated using the STANDARDIZE function. The formula `=STANDARDIZE(1012,1000,10)` will return 1.2 and `=STANDARDIZE(986,1000,10)` will return −1.4. Note that the first argument for STANDARDIZE is x, the second is μ and the third is σ.

Since the sum of all deviation from the mean is zero for all data sets, the average z-score is always zero. The variance of all z-scores is always 1, because it works out to σ^2/σ^2.

1.4 Histogram

When a numerical data set is large, it is convenient to group the data points into classes by choosing suitable *class intervals*. The class intervals must be chosen so as to have neither too few nor too many classes. The number of classes is usually kept between 5 and 20. The number of data points in a class is called the *frequency* of that class. A data set so grouped can then be represented by a *histogram.*

Definition 1.4.1: A *histogram* of a grouped data set is a bar chart where each bar represents the frequency of the corresponding class.

1.4.1 Creating Histograms from Raw Data

You can use the sheet named Raw Data in the workbook named Histogram to create a histogram from raw data. The template is shown in Figure 1.4.1. To use this template, enter the data in the shaded area of column B. Then enter the three values required for the horizontal axis, namely, start, interval width and end values. You may experiment with these three values and choose the combination that describes the distribution the best.

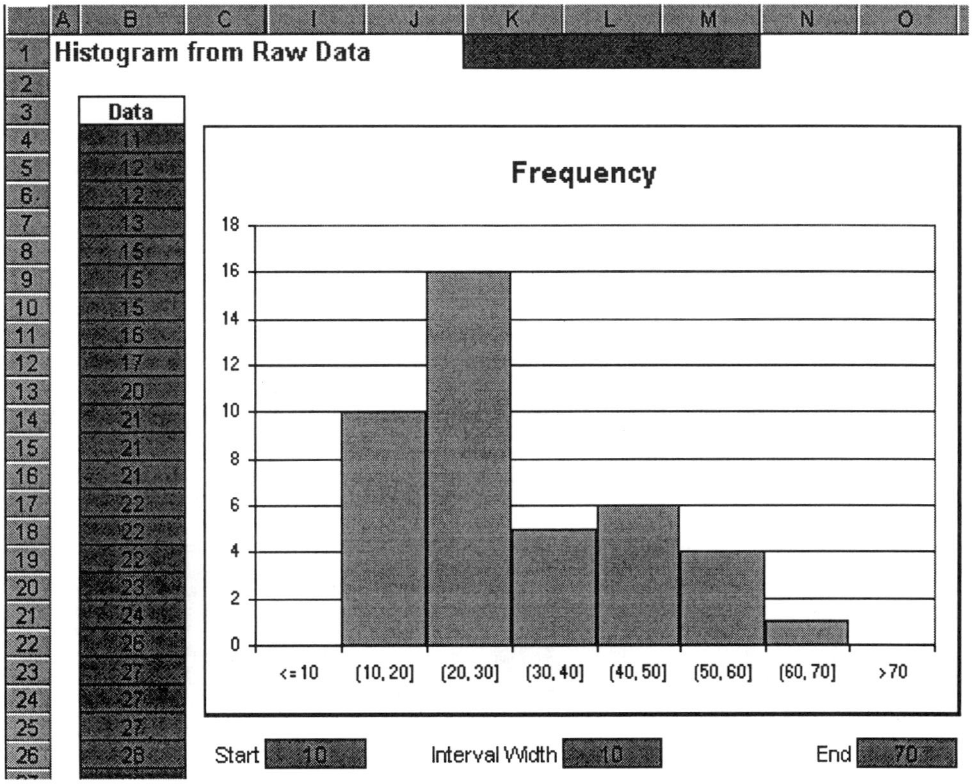

Figure 1.4.1. Drawing a histogram from raw data
[Raw Data in Histogram]

In the spreadsheet, there is also a histogram of the relative frequencies (not seen in the figure) to the right of the frequency histogram. The two histograms are the same except for the scale on the vertical axis. You may copy any of this histogram and embed it in another document such as a Microsoft Word document.

1.4.2 Creating Histograms from Grouped Data

At times, the data may not be raw but may be already grouped into classes. In this case the spreadsheet shown in Figure 1.4.2 can be used to draw histograms, frequency polygons and ogives. The data to be input are the class intervals and the corresponding frequencies.

Figure 1.4.2 shows only the histogram. In the spreadsheet there are relative frequency histogram, frequency polygon, relative frequency polygon and ogive. Any of these charts can be copied and embedded into other documents.

If you have to draw a frequency polygon or an ogive from raw data, first use the sheet named Raw Data (seen in Figure 1.4.1) to determine the class intervals and frequencies. Then transcribe the class intervals and frequencies onto the Grouped Data sheet (seen in Figure 1.4.2). The required charts are now created on this second sheet.

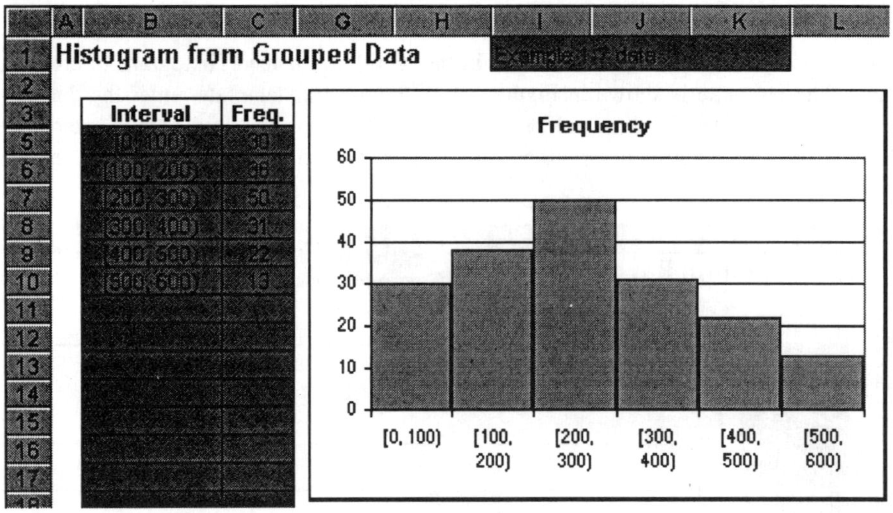

Figure 1.4.2. Histogram from Grouped Data
[Grouped Data in Histogram]

1.5 Skewness and Kurtosis

Having seen histograms, we can next consider the shape of histograms. In Figure 1.5.1 below,

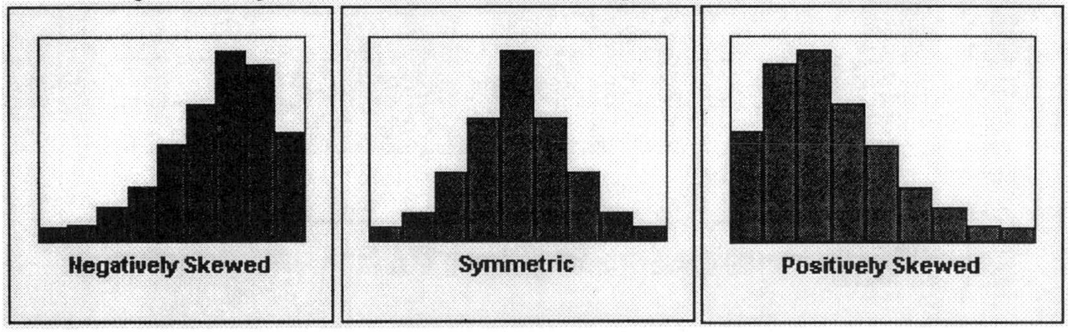

Figure 1.5.1 Same Mean and Variance but different Skewness

we see three histograms all of which have the same mean and variance. But they have very different shapes. The middle histogram is *symmetric* about a vertical axis at the center. The one on the left has a long tail to the left. Such a histogram is said to be *negatively skewed.* Similarly, the one on the right has a long tail to the right, and is said to be *positively skewed.* If we looked only at means and variances we would not notice these differences, and hence we use *skewness* to distinguish between the shapes.

Definition 1.5.1: *Skewness* of a data set is the average of the cubes of all the *z*-scores.

The cube of a negative number is negative. If a data set is symmetric, then for every positive *z*-score there will be a corresponding negative one and their cubes will cancel each other when summed. Thus the skewness of a symmetric data set is zero. In a negatively skewed data set, the negative cubes will outweigh the positives and we will end up with a negative value for skewness. By the same reasoning, the skewness of a positively skewed data set will be positive.

Skewness can be calculated on a spreadsheet using the function SKEW. In Figure 1.1.1, cell H13 has the formula `=SKEW(Data)` and calculates the skewness as –0.3515. Thus, our data is negatively skewed which means that it has some relatively very low values.

Because the true mean μ is not known and $\bar{x}$ is used in its place for calculating the deviations, there will be a downward bias in the calculation of sample skewness. To compensate for this bias, the skewness is multiplied by $n^2/(n-1)(n-2)$. While calculating skewness, Excel *always* assumes that the data is of a sample and carries out this bias correction. If the data is of a population, then the result from SKEW formula would be slightly wrong. The correct population skewness has been calculated in cell I13 in Figure 1.1.1. using a different formula.

It is possible that two or more different distributions have the same mean, variance and skewness but still have significantly different shapes. That is indeed the case for the three histograms in Figure 1.5.2. (Can you guess their common skewness?) To distinguish between them, another shape parameter called the *kurtosis* is defined. Kurtosis is a measure of peakedness of a histogram. Larger the kurtosis, more peaked is the shape.

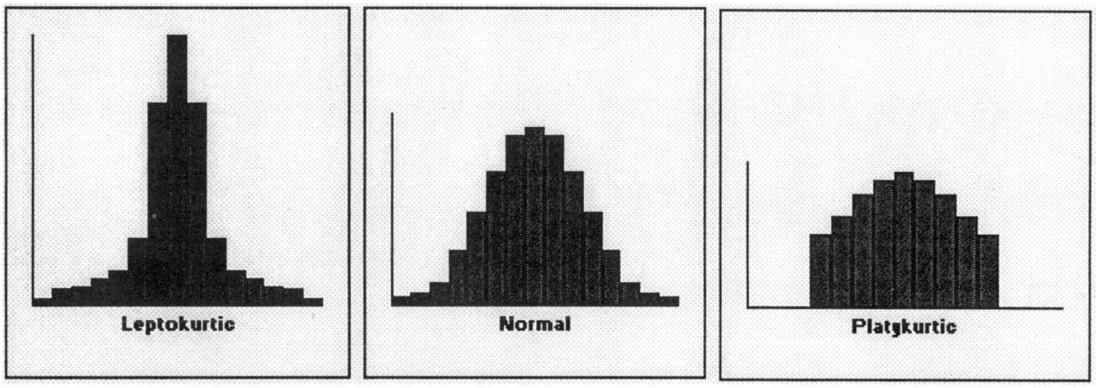

Figure 1.5.2 Same Mean, Variance and Skewness but different Kurtosis

Definition 1.5.2: The (absolute) *kurtosis* of a data set is the average of the fourth powers of the z-scores.

Since the fourth power of a negative number is positive, the (absolute) kurtosis is always positive. The middle histogram in Figure 1.5.2 is the famous Normal Distribution about which we will study a lot later. Its (absolute) kurtosis is 3. Many statisticians find it convenient to measure kurtosis relative to that of the Normal Distribution. Therefore, they subtract 3 from the average fourth power of the z-scores and call the result the (relative) kurtosis. The (relative) kurtosis of the Normal Distribution is thus zero. A shape that is more peaked than normal will have a positive (relative) kurtosis, and will be called *leptokurtic*. A shape that is less peaked than normal will have a negative (relative) kurtosis, and will be called *platykurtic*.

On a spreadsheet, the (relative) kurtosis of a data set can be calculated using the KURT function. In Figure 1.1.1, cell H14 has the formula

```
=KURT(Data)
```

and calculates the (relative) kurtosis as 0.1156. Thus, our data is slightly more peaked than the Normal Distribution.

If the data is of a sample, there would be the bias again in the calculation of sample kurtosis. While calculating kurtosis, Excel *always* assumes the data to be of a sample and applies a bias correction. If the data is of a population, the result from the KURT formula would be slightly wrong. The correct (relative) population kurtosis is calculated in cell I14 in Figure 1.2.1. using a different formula.

When one looks beyond kurtosis, it is true that two data sets may have the same mean, variance, skewness and kurtosis and yet be different. One can then calculate the average fifth power of the *z*-scores to distinguish between them. And after that, we may look at the average sixth power of the *z*-scores and so on. Theoretically, there is no end. But in practice, we rarely need to go beyond the fourth power, which is the kurtosis. If two data sets have the same mean, variance, skewness and kurtosis, then for most practical purposes, they would not be significantly different.

1.6 Box Plots

A type of chart that is common in descriptive statistics, is the *box plot*. A box plot is shown in Figure 1.6.1 with its parts labeled. The plot contains enough data for the user to visualize the shape of the distribution. Compared to other types of charts, a box plot is easier to draw manually. Moreover, it shows outliers and suspected outliers explicitly, while a histogram does not.

In the template Descriptive Statistics, the sheet marked Box Plot can be used to construct box plots from raw data. Just enter the data in column K of the sheet.

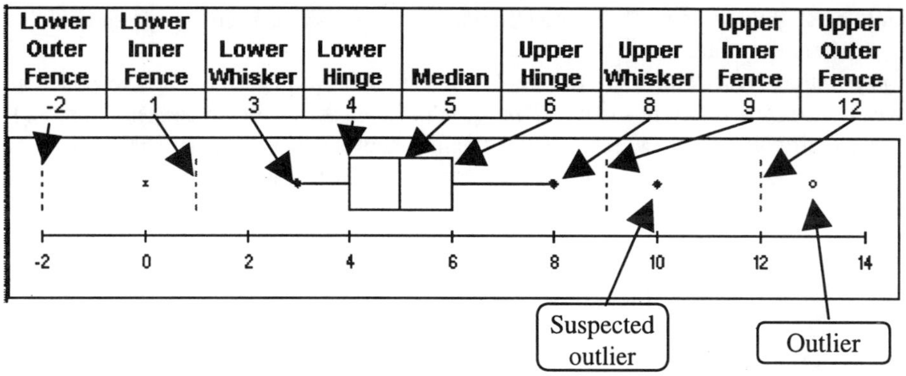

Figure 1.6.1. A Box Plot

1.7 Exercises

1. Do the exercise 1-62 to 66 from Aczel's book.

1.8 Projects

1. In this project we shall see how the statistics calculated for a data change when the data is modified by multiplying every data point by a number and/or adding another number to it.

(a) Open the workbook Descriptive Statistics and select the sheet named Project 1.
(b) Look at the formula in cell B5 to confirm how the data gets modified when the numbers in cells A2 and B2 are changed .
(c) Look at the formulas in the range E5:F6 to see how the two statistics are calculated.
(d) Enter similar formulas in the range E7:F16 to calculate the rest.
(e) Set **Mult** = 1 and **Add** = 5. The result is that 5 is added to every data point. Observe what has happened to all the statistics. Keeping **Mult** = 1 and changing **Add** to different values, observe how the statistics change. Make a formal statement such as, "If we add x to all the data points then the average would increase by x," for each of the statistics.
(f) Add an explanation for each statement made in 5. above. For the average, this will be, "If we add x to all the data points, then the sum of all the numbers will increase by $x*n$ where n is the number of data points. The sum is divided by n to get the average. So the average will increase by x."
(g) Do the same for multiplying all the data points by some number. This would require setting **Mult** equal to desired values and **Add** = 0.
(h) Do the same for multiplying and adding at once. This would require setting both **Mult** and **Add** to desired values.

2. Open the sheet named Project 2 in Descriptive Statistics. This sheet is also seen in Figure 1.8.1. It shows the calculation of the deviations, z-scores and z^k values for the data set in column B. The column average of z^k gives the k-th moment of the z-scores. The first seven moments of the z-scores have been tabulated on the right. Answer the following questions.

i) Why is the first moment zero? Will it be zero for all data sets?
ii) Why is the second moment 1? Will it be 1 for all data sets?
iii) The first, third, fifth and seventh moments are all zero for this data. Is that coincidence, or will all odd moments be zero for all data sets? When will all the odd moments of a data set be zero?
iv) Can you think of a data set for which all moments are zero?

	A	B	C	D	E	F	G	H	I
1	Moments of z								
2				Stdevp	k				
3				1.8708	2				
4		Data	Deviation	z-Score	z^k		k	k-th moment of z	
5		3	-1	-0.535	0.286				
6		2	-2	-1.069	1.143		1	0	Mean of z
7		5	1	0.535	0.286		2	1	Variance of z
8		4	0	0.000	0.000		3	0	Skewness of population
9		6	2	1.069	1.143		4	2	(absolute)Kurtosis of population
10		1	-3	-1.604	2.571		5	0	
11		4	0	0.000	0.000		6	4.629738	
12		7	3	1.604	2.571		7	0	
13	Column								
14	Average	4	0	0	1				

Figure 1.8.1 Moments of *z*
[Project 2 in Descriptive Statistics]

3. Whenever we calculated the deviations (to calculate *z*-scores, for example), we took the deviations from the mean. At times, we need to consider the deviations from another target value rather than the mean. Consider the case of a machine that bottles cola into 2 liter (2000 cc) bottles. The target is thus 2000 cc. The machine, however, may be bottling, say, 2004 cc on the average into every bottle. Call this 2004 cc the *process mean.* The damage from process errors is determined by the deviations of the data points from the target rather than from the process mean. The variance, though, is calculated with deviations from the process mean. Suppose we want to calculate the variance using deviations from the target value rather than the deviations from the process mean. Variance, as you may be aware, is the average of the squared deviations. Let "SSD(Target)" denote the sum of the squared deviations from the target. [For example SSD(2000) denotes the summ of squared deviations when the deviations are taken from 2000.] Dividing the SSD by the number of data points gives the Average SSD(Target).

The spreadsheet seen in Figure 1.8.2 is set up to calculate the deviations from the target, SSD(Target) and the Average SSD(Target). This spreadsheet is the sheet named Project 3 in the workbook named Descriptive Statistics. As seen in the figure, Column B contains the data, which has a process mean of 2004. [Strictly speaking, this would be sample data. But to simplify matters, let us assume that this is population data.] Note that the population variance (VARP) is 3.5 and the Average SSD(2000) is 19.5.

In the range G5:H13, a table has been created to see the effect of changing the Target on Average SSD(Target). The offset refers to the difference between the Target and the process mean.

i) Study the table and find an equation that relates the Average SSd to VARP and the Offset. [Hint: Note that while calculating SSD, the deviations are squared. So, think in squares.]

ii) Using the equation you found in i) above, prove that the Average SSD(Target) is minimized when the Target equals the process mean. Verify this using the Solver tool. [Hint: Make the Solver minimize the Average SSD in cell D14 by changing the Target in cell C1.]

	A	B	C	D	E	F	G	H	I
1		Target	2000						
2									
3		Data	Deviation from Target	Deviation^2		Offset	Target	Average SSd	
4		2003	3	9					
5		2002	2	4		-4	2000	19.5	
6		2005	5	25		-3	2001	12.5	
7		2004	4	16		-2	2002	7.5	
8		2006	6	36		-1	2003	4.5	
9		2001	1	1		0	2004	3.5	<- VARP
10		2004	4	16		1	2005	4.5	
11		2007	7	49		2	2006	7.5	
12						3	2007	12.5	
13	Mean	2004	SSd	156		4	2008	19.5	
14	VARP	3.5	Avg. SSd	19.5					

Figure 1.8.3. Average SSd from Target
[Project 3 in Descriptive Statistics]

2. Probability

2.1 Definitions

The probability of an event is a numerical measure of its likelihood. It is a number between 0 and 1, and larger the number, larger the likelihood. A probability of 0 represents that the event certainly will not occur and a probability of 1 that it certainly will.

Before we get to the concepts of probability we need to overview some concepts from set theory. We start with the concept of an *element*. If we are concerned with all the employees of a company, then each employee, and nothing else, is an element. If we are concerned with the number of defectives among 100 parts that are being inspected then each integer 0, 1, 2, ..., 100, and nothing else, is an element.

Definition 2.1.1: A *set* is a collection of elements.

We denote sets by uppercase letters *A, B* etc. We denote elements by lower case letters *x, y* etc. If *x* belongs to *A* we write $x \in A$. If *x* does not belong to *A*, we write $x \notin A$. Two sets are equal if they contain the same collection of elements.

Definition 2.1.2: The *universal set* is one that contains all the elements.

We denote the universal set by *S*. If *x* is any element then $x \in S$.

Definition 2.1.3: The *empty set* is one that contains no elements.

We denote the empty set by $\varnothing$. If x is any element then $x \notin \varnothing$.

Definition 2.1.4: If every element that belongs to *A* also belongs to *B*, then *A* is a *subset* of *B*.

If *A* is a subset of *B* then we write $A \subseteq B$. We note that if *A* is any set, $\varnothing \subseteq A \subseteq S$.

Definition 2.1.5: The *intersection* of two sets *A* and *B* is the set of all elements that belong to both A and B.

We denote the intersection of *A* and *B* by $A \cap B$.

Definition 2.1.6: The *union* of two sets *A* and *B* is the set of all elements that belong to *A* or *B* or both.

We denote the union of *A* and *B* by $A \cup B$. Note that $A \cap B$ denotes elements in *A* and *B* at the same time, and $A \cup B$ denotes elements in *A* or *B*.

Definition 2.1.7: The *complement* of a set *A* is the set of all elements that do not belong to *A*.

We denote the complement of *A* by $\overline{A}$. If *A* is any set then $A \cap A = A; A \cup A = A;$ $A \cap \varnothing = \varnothing; A \cup \varnothing = A; A \cap S = A; A \cup S = S; \overline{\varnothing} = S; \overline{S} = \varnothing;$ $A \cap \overline{A} = \varnothing; A \cup \overline{A} = S$.

Definition 2.1.8: Two sets *A* and *B* are said to be *disjoint* if they have no element in common.

We shall now relate these set theory concepts to probability theory. Suppose an *experiment* may result in any one of several *outcomes*. The outcomes then become the elements. In probability theory, elements are referred to as *sample points;* the universal set is referred to as the *sample space;* and a set of sample points is referred to as an *event*. We say that an event has occurred if any one of the sample points that belong to it occurs. We are now ready to define the probability of an event.

Definition 2.1.9: If every sample point in a finite sample space is equally likely, then the probability of an event A is the number of sample points that belong to A divided by the number of sample points in *S*.

Let P(*A*) denote the probability of event *A*, *n*(*A*) the number of sample points in *A* and n(*S*) the number of sample points in *S*. We can then write the formula for P(*A*) as

$$P(A) = n(A)/n(S). \qquad (2.1.1)$$

Suppose we toss a die. There are six sample points 1, 2, ..., 6 each of which is equally likely. Let *A* denote the event where an odd number shows up. Event *A* then contains three sample points 1, 3, and 5. The probability of *A* is therefore 3/6 = 1/2.

Suppose we pick a card at random from a standard deck of playing cards. There are 52 sample points in the sample space, each one equally likely. Let *A* denote the event we pick an ace. Then event *A* contains 4 sample points and therefore the probability of *A* is 4/52 = 1/13.

While using Formula 2.1.1, one should be careful about the following conditions that must be satisfied.

1. All the sample points should have been identified without any omission.
2. The sample points should be *mutually exclusive*, i.e., if one of them occurs, no other can also occur.
3. All the sample points should be equally likely.

The third condition of equal likelihood can be difficult to ensure. Consider the following problem: We toss a coin twice. What is the probability that they are both heads?

One may list the sample points as:

1. The tosses are both heads
2. The tosses are not both heads

In this case, the answer turns out to be 1/2.

Another way to list the sample points is:

1. We get 0 heads
2. We get 1 heads
3. We get 2 heads.

In this case, the answer turns out to be 1/3.

Yet another way is:

1. First toss heads; second toss heads
2. First toss heads; second toss tails
3. First toss tails; second toss heads
4. First toss tails; second toss tails

In this case, the answer turns out to be 1/4. Which answer is correct? Only the last one is correct, because only there the sample points are all equally likely.

Often, we need to calculate the probability of the intersection or union of events. Formula 2.1.1 can be extended for this purpose. We have

$$P(A \cap B) = n(A \cap B)/n(S) \qquad (2.1.2)$$

$P(A \cap B)$ is also known as the *joint probability* of events A and B.

$$P(A \cup B) = n(A \cup B)/n(S). \qquad (2.1.3)$$

Noting that $n(\overline{A}) = n(S) - n(A)$ we can write $P(\overline{A}) = [n(S) - n(A)]/n(S)$ which gives

$$P(\overline{A}) = 1 - P(A) \qquad (2.1.4)$$

Consider $n(A \cup B)$ in Equation 2.1.3. Can we replace it with $[n(A) + n(B)]$? No. If A and B have some common elements, then they will be counted twice in $[n(A) + n(B)]$. To avoid this double counting, we can subtract $n(A \cap B)$ from $[n(A) + n(B)]$ and get

$$n(A \cup B) = n(A) + n(B) - n(A \cap B).$$

Equation 2.1.3 can then be rewritten as

$$P(A \cup B) = P(A) + P(B) - P(A \cap B) \qquad (2.1.5)$$

This Equation 2.1.5 is very useful since most of the time the right hand side probabilities are known and we will need to calculate $P(A \cup B)$. This equation is also known as the *Addition Rule* for probabilities.

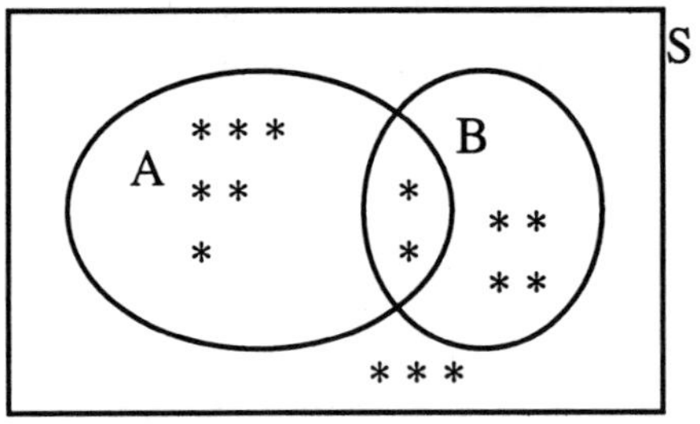

Figure 2.1.1. A Venn diagram

Let us apply the above formulas to the case depicted in the Venn diagram of Figure 2.1.1. We see the sample space *S* with two events *A* and *B*. Let each of the *'s represent a sample point, and assume that they are all equally likely. Now, $n(A) = 8$ and $n(S) = 15$. Hence, $P(A) = 8/15$. Similarly, $P(B) = 6/15$; $P(A \cap B) = 2/15$; $P(A \cup B) = 12/15$. Note that the equations 2.1.4 and 2.1.5 are satisfied.

Next we shall see an important concept called *conditional probability*. Refer to the Venn diagram in Figure 2.1.1. When we are calculating $P(A)$, let us say we are told event *B* has occurred. In other words, the actual outcome is one of the *'s inside *B*. Can we still say $P(A) = 8/15$? No. The sample points we can consider now are only those inside *B*. In effect, *B* has become our new sample space. Within this new sample space which contains 6 sample points, 2 belong to *A*. Hence, $P(A)$ given that event *B* has occurred is 2/6. We call this the *conditional probability of A given B* and denote it by P(A | B). The event written after the vertical line | is the condition. We shall write this as our next formula.

$$P(A \mid B) = n(A \cap B)/n(B) \qquad (2.1.6)$$

Combining the formulas 2.1.1, 2.1.2 and 2.1.6, we can write

$$P(A \mid B) = P(A \cap B)/P(B) \qquad (2.1.7)$$

Formula 2.1.7 is the famous *Bayes' Rule* named after an 18th century English clergyman who wrote much about this rule. This formula can be re-written with P($A \cap B$) on the left hand side:

$$P(A \cap B) = P(B)*P(A \mid B) \qquad (2.1.8a)$$

Or, we may interchange A and B and write the above formula as:

$$P(A \cap B) = P(A)*P(B \mid A) \qquad (2.1.8b)$$

The Formula 2.1.8 (a or b) is known as the *Multiplication Rule* of probabilities.

At times P(A) and P(A | B) may be equal, which would imply that the occurrence of event B did not in any way influence the probability of event A. We then say A and B are *independent events*.

Definition 2.1.10: Two events A and B are said to be *independent* if P(A) = P(A | B).

As an example, consider the experiment of selecting one adult at random from a community. The sample space S then contains all the adults in the community. Let A denote those with lung disease and B denote those who smoke. If P(A) is found to be equal to P(A | B), then we have reason to believe that smoking does not influence the chances of one getting a lung disease. We would then declare that smoking and getting a lung disease are independent. If the opposite is true, we would then declare them *dependent*. Thus, independence among events is an important practical concept.

Using algebra, it can be shown that if A and B are independent events then the pairs A & $\overline{B}$, $\overline{A}$ & B, $\overline{A}$ & $\overline{B}$ are all independent pairs.

In Figure 2.1.1, we see that P(A) = 8/15 and P(A | B) = 1/3. Hence A and B are not independent events.

An important consequence of independence is that the joint probability of two independent events will be equal to the product of their individual probabilities. Looking at the definition of independence and Formula 2.1.8a, we can write

$$\text{when } A \text{ and } B \text{ are independent } \; P(A \cap B) = P(A)*P(B). \qquad (2.1.9)$$

Formula 2.1.9 extends to any number of independent events:

$$P(A_1 \cap A_2 \cap \ldots \cap A_n) = P(A_1)*P(A_2)*\ldots*P(A_n).$$

The above formula considerably simplifies some probability calculations. Consider the case where we need the probability of at least one success from many independent trials. If A_i denotes success in the i-th trial, then we need the probability of the union of all A_i. To carry out this computation, first we write using Formula 2.1.4,

$$P(A_1 \cup A_2 \cup \ldots \cup A_n) = 1 - P(\overline{A_1 \cup A_2 \cup \ldots \cup A_n})$$

We then use *De Morgan's rule* to get:

$$\overline{A_1 \cup A_2 \cup \ldots \cup A_n} = \overline{A_1} \cap \overline{A_2} \cap \ldots \cap \overline{A_n}. \qquad (2.1.10)$$

Now since the events are independent

$$P(\overline{A_1} \cap \overline{A_2} \cap \ldots \cap \overline{A_n}) = P(\overline{A_1}) * P(\overline{A_2}) * \ldots * P(\overline{A_n}).$$

Combining all we get,

$$P(A_1 \cup A_2 \cup \ldots \cup A_n) = 1 - P(\overline{A_1}) * P(\overline{A_2}) * \ldots * P(\overline{A_n}). \qquad (2.1.11)$$

As an example, consider the following problem: A graduating student applies for 5 different jobs. The probability that she gets an offer at each of the five applications are, respectively, 0.2, 0.1, 0.6, 04 and 0.2. Further, the chances at one job does not affect the chances at any other, meaning that the events are all independent. What is the probability that she will get at least one offer?

Let A_1 through A_5 denote the event of getting an offer at each of the five applications. The probability of getting at least one offer is $P(A_1 \cup A_2 \cup \ldots \cup A_5) = 1 - P(\overline{A_1})*P(\overline{A_2})*\ldots*P(\overline{A_5}) = 1 - 0.8*0.9*0.4*0.6*0.8 = 1 - 0.1382 = 0.8618$. Thus, there is about 86% chance that she would get at least one offer.

Let us look at a spreadsheet template that solves this and similar problems. In the workbook named Probbility, open the sheet named SuccessProbs. On this sheet, the range A4:A8 has the success probabilities of each trial from the above problem. In cell C4, the *array formula*

```
=1-product(1-SuccessProbs)
```

has been entered by pressing `CTRL+SHFT+Enter`. The answer 0.8618 appears in this cell.

2.2 Combinatorial Concepts

The calculation of probabilities often needs counting large numbers of sample points. The counting can be simplified through formulas. Suppose we need to go from city A to city C via city B. Let there be 2 different roads from A to B and three from B to C. In how many different ways can we go from A to C? Each of the two roads from A to B can be combined with each of the three roads from B to C and therefore there are 2 * 3 = 6 different ways. Going further, if there are 4 different roads from C to D, then we can go from A to D in 2 * 3 * 4 = 24 different ways. Generalizing this idea, we have a *counting rule*:

> If there are n tasks to do and task i can be done in N_i different ways then the total number of different ways in which all the tasks can be done is $N_1 * N_2 * \ldots * N_n$.

2.2.1 Permutation

Suppose there are five different objects and three different boxes, and that we need to put exactly one object in each of the three boxes. In how many ways can we do this? Let ${}_n\mathbf{P}_r$ denote the number of different ways that n objects can be permuted in r boxes. The question is what is ${}_5\mathbf{P}_3$?

We note that the first box can be filled five different ways since it can be filled with any one of the five objects. After the first box is filled, there are only four objects left and therefore the second box can be filled only four different ways. Then the third box can be filled three different ways. Using the counting rule above, the three boxes together can be filled 5 * 4 * 3 = 60 different ways. As another example, with 10 objects and 4 boxes, there will be 10 * 9 * 8 * 7 different ways.

If there were 100 objects and 76 boxes, the answer is 100 * 99 * ... * 25, which we would like to write in a more compact way. The *factorial* notation is useful here. The notation 10!, read ten factorial, denotes 10 * 9 * 8 * ... * 1. In General, $n! = n * (n\text{-}1) * (n\text{-}2) * \ldots * 1$. We note that the expression 10 * 9 * 8 * 7 can be written 10!/6!, and coming to the one we started with, 100 * 99 * ... * 25 can be written 100!/24!. Thus, we could say ${}_{100}\mathbf{P}_{76}$ = 100!/(100-76)!. Generalizing this result, we get the permutation formula:

$$ {}_n\mathbf{P}_r = n!/(n - r)! \qquad (2.2.1)$$

The counting rule reveals that ${}_n\mathbf{P}_n$ should be $n!$; according to the formula above it is $n!/0!$. No discrepancy would be there if 0! is declared as 1, and that is indeed the case. By definition, 0! = 1.

On a spreadsheet, ${}_n\mathbf{P}_r$ can be calculated using the function `=PERMUT(n,r)`. For instance, `=PERMUT(5,3)` will return 60.

2.2.2 Combination

Suppose we have to form a committee of 3 members drawing them from a pool of 5 candidates labeled A, B, C, D and E. In how many different ways can we do this? It is important to realize that the answer is not ${}_5\mathbf{P}_3$ because A B C and A C B are the same committee, although different permutations. Let us look at the notation for combinations. The number of different ways in which a combination of r objects can be selected from a pool of n objects is denoted by $\binom{n}{r}$. In some books, it is denoted by ${}_n\mathbf{C}_r$.

The *order* matters in permutations but not in combinations. Thus, when we count permutations, we will count the 6 permutations A B C, A C B, B A C, B C A, C A B, and C B A as distinct, but they are all the same committee and must be counted as only 1 for combinations. In other words, a committee of three, having 3! possible permutations of members, will be counted 3! times in permutations. If the committee had r members, it would be counted $r!$ times in permutations. Since each of the $\binom{n}{r}$ committees will be counted $r!$ times in permutations we have, ${}_n\mathbf{P}_r = r! * \binom{n}{r}$.

We can rewrite this to get a formula for combinations:

$$\binom{n}{r} = \frac{n!}{r!(n-r)!} \tag{2.2.2}$$

On a spreadsheet, we can calculate $\binom{n}{r}$ using the formula `=COMBIN(n,r)`. For instance, the formula `=COMBIN(5,3)` returns 10.

2.3 Cross Tabulation and Joint Probability Table

Often a cross tabulation of frequency counts forms the basis for probability calculations. In Figure 2.3.1, we see a cross tabulation of parts made by three machines classified as good and defective. The Totals row and column have been created using the SUM function. For instance, cell B5 has the formula `=SUM(B3:B4)`. [This tabulation is not very different from a Venn diagram. One can treat the table (excluding the Total row and the Total column) as the sample space, each cell as an event, and the number in each cell as representing the number of sample points (*'s) belonging to that event.] This spreadsheet is available in the workbook named Probability, as the sheet named Joint Prob Table.

If we pick a part at random, what is the probability that it was made by Machine 1? The answer is 1269/5946 = 0.2134. What is the probability that it is made by Machine 1 <u>and</u> that it is good? The answer is 1245/5946 = 0.2094. If we create a Joint Probability Table, then these probabilities can be read off directly. To create the table, we divide all the numbers by the grand total of 5946.

	A	B	C	D	E
1	**Cross Tabulation**				
2		Machine 1	Machine 2	Machine 3	*Total*
3	Good	1245	2480	2144	5869
4	Defective	24	35	18	77
5	*Total*	1269	2515	2162	5946
7	**Joint Probability Table**				
8		Machine 1	Machine 2	Machine 3	*Total*
9	Good	0.2094	0.4171	0.3606	0.9871
10	Defective	0.0040	0.0059	0.0030	0.0129
11	*Total*	0.2134	0.4230	0.3636	1.0000
12					
13	P(Machine 1|Good) =		0.2121		

Figure 2.3.1 Cross Tabulation and Joint Probability Table
[Joint Prob Table in Probability]

In the spreadsheet in Figure 2.3.1, a Joint Probability Table has been created in the range A8:E11. It can be created quickly by doing the following:

- Select the range A2:E5 and click on the copy icon.
- Click on cell A8 and click on the paste icon.
- Select the range B9:D10
- Enter the formula `=B3:D4/E5` and Press `CTRL+SHFT+Enter.`

It is possible to calculate conditional probabilities using either the cross tabulation or the Joint Probability Table. For instance, what is the probability that a randomly selected part was made by Machine 1, given that it is good? We need P(Machine 1|Good) = 0.2094/0.09871 = 1245/5869 = 0.2121. It has been calculated in cell C13.

2.4 Exercises

1. Solve problems 2-58, 2-59 from Aczel's book.
2. Solve problem 2-101, 2-102 in Aczel's book.
3. Use the SuccessProbs sheet in the workbook named Probability and solve Problem 2-99 in Aczel's book. [*Hint*: *Create a sufficiently long list of SuccessProbs using a suitable formula.*]
4. In the figure below, each the *'s are mutually exclusive, exhaustive and equally probable. There are a total of 16 *'s in the sample space *S*. Calculate the following probabilities.

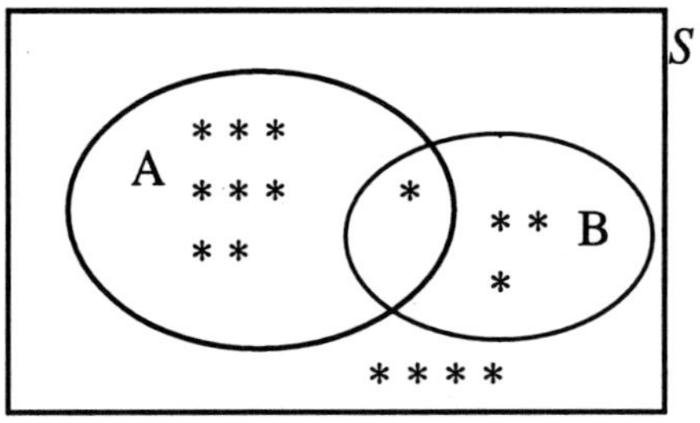

i. $P(\overline{A})$
ii. $P(\overline{A} \mid B)$
iii. $P(A \cup \overline{B})$
iv. $P(A \cap \overline{B} \mid A \cup \overline{B})$

3. Random Variables

3.1 Introduction

Many quantities, especially those about the future, are not known for sure. The outside temperature tomorrow, the price of a stock next week, the number of highway accidents during next month and the breaking strength of a bolt are all uncertain. But we have to deal with uncertain quantities in practice and may even have to make serious decisions based on them. Hence we study them. Uncertain quantities are called *random variables*.

Definition 3.1.1: A *random variable* is an uncertain quantity whose value depends on chance.

Some quantities such as length, time, weight and speed can vary continuously and are therefore *continuous*. A length can be 16.204596... yards, and is continuous. But if we count the number of employees in a company, the quantity is *discrete*. It can be 2048 of 2049 but not 2048.350876... . A random variable can be either *continuous* or *discrete*.

Definition 3.1.2: A random variable is *discrete* if it can assume at most a countable number of possible values. If it can assume an uncountable number of possible values then it is *continuous*.

When a random variable is discrete, it can be described fully by tabulating all of its possible values and the corresponding probabilities. The sheet shown in Figure 3.1.1, describes the random variable X, which can assume any one of the values listed under column $\boldsymbol{x}$, namely, 0, 1, 2, 3, 4, 5. The **P(*x*)** column contains the corresponding probabilities. That is, $P(X = 0) = 0.1$, $P(X = 1) = 0.2$ and so on. The probabilities should add to 1.

Another useful probability is the *cumulative probability* F(x) defined as

$$F(x) = P(X \leq x).$$

It can be used to calculate the probability that X will be between two given values, because

$$P(x_1 < X \leq x_2) = F(x_2) - F(x_1).$$

The function PROB is also useful here. Referring to Figure 3.1.1, the formula

```
=PROB(A4:A9,B4:B9,2)
```

will return $P(X = 2)$ which is 0.3. The formula

```
=PROB(A4:A9,B4:B9,2,4)
```

will return $P(2 \leq X \leq 4)$ which is 0.6.

The cumulative probabilities have been calculated on the spreadsheet in the range C4:C9. To do it,

- In cell C4 enter the formula `=B4`
- In cell C5 enter the formula `=C4+B5`

Copy the formula from C5 into the range C6:C9.

Table 3-4, Aczel 3rd Edn., p. 104

x	P(x)	F(x)
0	0.1	0.1
1	0.2	0.3
2	0.3	0.6
3	0.2	0.8
4	0.1	0.9
5	0.1	1
Total	1	

Figure 3.1.1 Discrete Probability Table
[Discrete in Distributions]

Although the exact value of a random value is not known for sure, what would be its good *representative* value? It could be the mean, the median or the mode.

Definition 3.1.3: The mean μ of a random variable is called its *expected value*, E[X], and is defined as

$$\mu = E[X] = \sum_{all x} x * P(x) \tag{3.1.1}$$

Definition 3.1.4: The *median* of a random variable X is that value x for which $F(x) = 0.5$.

Definition 3.1.5: The *mode* of a random variable X is that value x for which $P(x)$ is the maximum. Sometimes we need the expected value of a function of X, say $h(X)$, defined as

$$E[h(X)] = \sum_{all x} h(x) * P(x) \qquad (3.1.2)$$

In the sheet named E[h(x)] in the workbook named Distributions, the expected value of $h(X)$ where

$$h(X) = 2*X - 8000$$

has been calculated. See Figure 3.1.2. Cell B4 has the formula `=2*A4-8000`, cell D4 has `=B4*C4` and cell D9 has `=SUM(D4:D8)`. This calculation $E[h(X)]$ using two columns for $h(x)$ and $h(x)*P(x)$ is tedious. A short cut is to use the SUM array formula (See Chapter 0). In Figure 3.1.2, the range A4:A8 has been named `Sales` and the range C4:C8 has been named `Probs`. The cell D11 has the array formula `=SUM((2*Sales-8000)*Probs)` entered by pressing `CTRL+SHFT+Enter`, and calculates $E[h(X)]$ as 5400 in one stroke. We shall always use this short cut in future.

	A	B	C	D	E
1	Calculating E[h(X)]				
2					
3	x	h(x)	P(x)	h(x)*P(x)	
4	5,000	2,000	0.2	400	
5	6,000	4,000	0.3	1,200	
6	7,000	6,000	0.2	1,200	
7	8,000	8,000	0.2	1,600	
8	9,000	10,000	0.1	1,000	
9			E[h(X)] =	5,400	
10	Using SUM array formula:				
11			E[h(X)] =	5,400	{=SUM((2*Sales-8000)*Probs)}

Figure 3.1.2 Calculating $E[h(X)]$ using the SUM array formula [E[h(x)] in Distributions]

Definition 3.1.6: The *variance* of a random variable X, written $V(X)$ or σ^2, is the expected value of its squared deviation from the mean. Formally,

$$\sigma^2 = V(x) = E[(x - \mu)^2]. \qquad (3.1.3)$$

In other words, if $h(x)$ is defined as $h(x) = (x - \mu)^2$, then the variance is $E[h(X)]$. There is also another formula for $V(x)$ which is useful if $V(x)$ is to be calculated manually. It is:

$$V(x) = E[X^2] - \mu^2 \qquad (3.1.4)$$

The advantage of Formula (3.1.4) is that it does not require the calculation of deviations; the disadvantage is that it requires the calculation of $E[X^2]$ which can be tedious when the x-values are large. We shall next see the definitions of descriptive statistics of a random variable.

Definition 3.1.7: The *standard deviation* of a random variable, written $SD(X)$ or σ, is the positive square root of its variance.

Definition 3.1.8: The *skewness* of a random variable is the expected value of z^3 where $z = (x - \mu)/\sigma$.

Definition 3.1.9: The (absolute) *kurtosis* of a random variable is the expected value of z^4.

Figure 3.1.3 shows the sheet named Descriptive Stats in the workbook named Distributions. The data has been entered in columns A and B. The x-values have been named `X` and the probabilities have been named `P`. The descriptive statistics have been calculated in column E and the formulas used, mostly SUM array formulas (see Chapter 0), have also been shown right next to each statistic.

This sheet can be used to calculate the descriptive statistics of any random variable. Just enter the x-values and the probabilities in columns A and B. Make sure that the sum of all the probabilities shown in cell B2 equals 1.

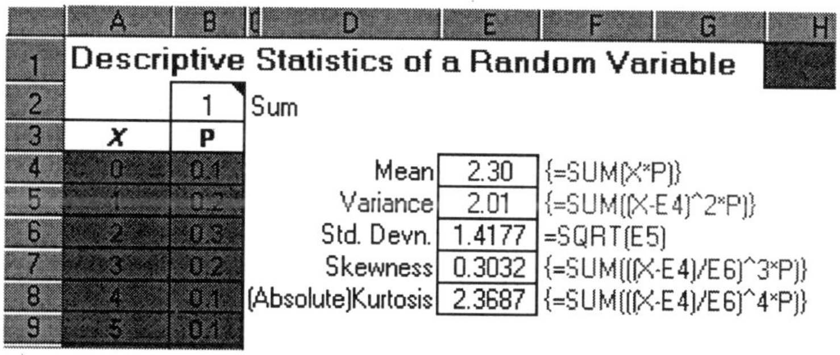

Figure 3.1.3 Descriptive Statistics of a Random Variable
[Descriptive Stats in Distributions]

3.1.1 Sum of Two Random Variables

Sometimes we have two random variables X and Y, and need the descriptive statistics of the sum of the two. The following formulas are then useful.

$$E[X + Y] = E[X] + E[Y] \tag{3.1.5}$$

If X and Y are independent, then we have

$$V(X + Y) = V(X) + V(Y) \tag{3.1.6}$$

3.1.2 The Use of Standard Random Variables

When the probability distribution of a discrete random variable is known fully, the statistical calculations we have seen so far are all possible. In practice, however, the probability distribution of a random variable may not be known fully. Empirically assessing the probability of every possible value of a random variable can be difficult, even impossible, especially when the probabilities are very small. But we may be able to find out what *type* of random variable the one at hand is by examining the causes that make it random. Knowing the type, we can often approximate the random variable to a *standard* one for which convenient formulas are available to calculate expected value, variance and other statistics. We shall study some popular standard random variables and look at related spreadsheet formulas. Side by side, we shall examine the conditions under which a random variable encountered in practice can be approximated to one of the standard ones.

3.2 Bernoulli Random Variable

The first standard random variable we shall study is the *Bernoulli random variable*, so named in honor of the mathematician, James Bernoulli (1654-1705). It is the building block for other random variables in this chapter. The distribution of a Bernoulli random variable X is given in the table below. As seen in the table, x is 1 with probability p and 0 with probability $(1-p)$. The case where $x = 1$ is called "success" and the case where $x = 0$ is called "failure."

Bernoulli Distribution

x	$P(x)$
1	p
0	$1-p$

Note that $E[X] = 1*p + 0*(1-p) = p$; $E[X^2] = 1^2*p + 0^2*(1-p) = p$; and therefore $V(X) = p - p^2 = p(1-p)$. Often the quantity $(1-p)$, which is the probability of failure, is denoted by the symbol q, and thus $V(X) = pq$. If X is a Bernoulli random variable with probability of success p, then we write $X \sim BER(p)$ where the symbol "~" is read "is distributed as" and BER stands for Bernoulli. The characteristics of a Bernoulli random variable are summarized in Box 3.2.1.

Box 3.2.1 Bernoulli Distribution

If $X \sim \text{BER}(p)$, then

$$P(1) = p;\ P(0) = 1-p$$
$$E[X] = p$$
$$V(X) = p(1-p)$$

Example: If $p = 0.8$, then

$$E[X] = 0.8$$
$$V(X) = 0.16.$$

Let us look at an instance of Bernoulli random variable. Suppose an operator uses a lathe to produce pins, and the lathe is not perfect in that it does not always produce a good pin. Rather, it has a probability p of producing a good pin and $(1-p)$ of producing a defective one. Just after the operator produces one pin, let X denote the "number of good pins produced." Clearly, X is 1 if the pin is good and 0 if it is defective. Thus, X follows exactly the distribution tabulated above, and therefore $X \sim \text{BER}(p)$. In general, if a trial is undertaken to achieve something, where the outcome can only be either success or failure, then the trial is a *Bernoulli trial.* The number of successes X in one Bernoulli trial, which can be 1 or 0, is a Bernoulli random variable.

A Bernoulli random variable is too simple to be of immediate practical use. But it forms the building block of the binomial random variable which is quite useful in practice. The binomial random variable in turn is the basis for many other useful cases.

3.3 The Binomial Random Variable

Consider n number of *identically and independently distributed* Bernoulli variables $X_1, X_2, \ldots, X_n$. Here, identically means that they all have the same p, and independently means that the value of one X does not in any way affect the value of another. For example, the value of X_2 does not affect the value of X_3 or X_5 and so on. This independence property is exploited later to obtain joint event probabilities by multiplying individual event probabilities. Such a *sequence* of identically and independently distributed Bernoulli variables is called a *Bernoulli Process.*

Suppose an operator produces n pins, one by one, on a lathe that has p probability of making a good pin at each trial. If this p remains constant throughout, then independence is guaranteed and the sequence of numbers (1 or 0) denoting the good and bad pins produced in each of the n trials is a Bernoulli process. For example, in the sequence of eight trials denoted by

$$0\,0\,1\,0\,1\,1\,0\,0$$

the third, fifth and sixth are good pins, or successes.

In practice, we are usually interested in the total number of good pins rather than the sequence of 1's and 0's. In the example above, three out of eight are good. In the general case, let X denote the total number of good pins produced in n trials. We then have

$$X = X_1 + X_2 + \ldots + X_n \quad \text{where all } X_i \sim \text{BER}(p) \text{ and are independent.}$$

Such an X, which is the sum of many Bernoulli random variables, is called a *Binomial Random Variable.*

3.3.1 Binomial Distribution Formulas

Consider the case where 5 trials are made, and in each trial the probability of success is 0.6. What is the probability that the number of successes in the 5 trials will be exactly 3?

First, we note that there are $\binom{5}{3}$ ways of getting three successes out of five trials. Next we observe that each of these possibilities has $0.6^3 * 0.4^2$ probability of occurrence. And therefore,

$$P(X = 3) = \binom{5}{3} * 0.6^3 * 0.4^2 = 0.3456.$$

We can now generalize the above formula where n is the number of trials and p is the probability of success. What is $P(X = x)$?.

$$P(X = x) = \binom{n}{x} p^x (1-p)^{(n-x)} \qquad \text{for } x = 0, 1, 2, \ldots, n \qquad (3.3.1)$$

Equation (3.3.1) is the famous binomial probability formula.

To describe a binomial random variable we need two parameters, namely, n and p. We write $X \sim B(n, p)$ to indicate that X is binomially distributed with n number of trials and p probability of success in each trial. The letter B in the expression stands for Binomial.

In Excel, the formula

```
=BINOMDIST(x,n,p,FALSE)
```
(3.3.2)

would return the probability $P(X = x)$ and

```
=BINOMDIST(x,n,p,TRUE)
```
(3.3.3)

would return the *cumulative probability* $P(X \leq x)$ which is the probability of at most x successes. Using these formulas, we can calculate any binomial probability. It would be better, however, if we could see many probabilities at once. Then we can solve problems more readily. The Distributions workbook contains one sheet for each of the distributions described in this chapter. The binomial distribution sheet is shown in Figure 3.3.1.

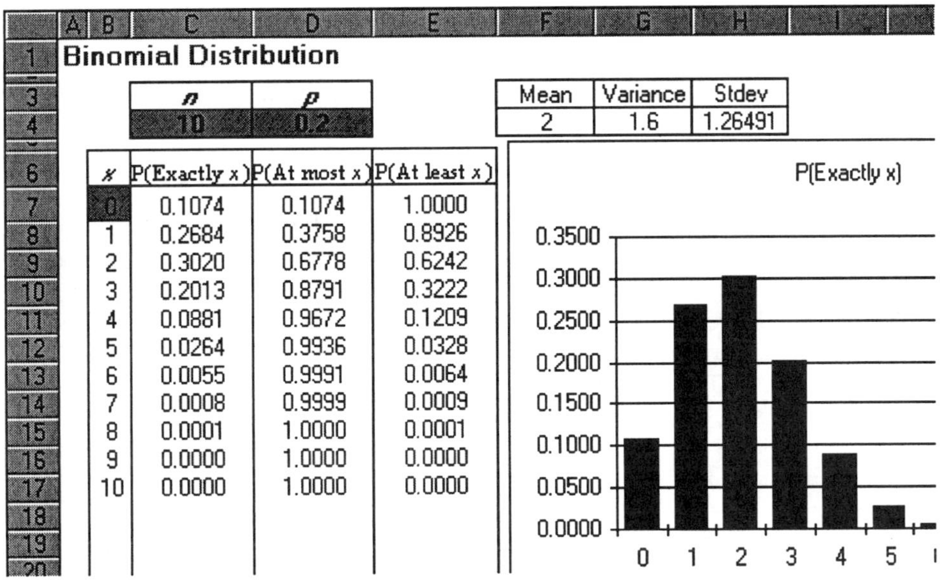

Binomial Distribution

n	p
10	0.2

Mean	Variance	Stdev
2	1.6	1.26491

x	P(Exactly x)	P(At most x)	P(At least x)
0	0.1074	0.1074	1.0000
1	0.2684	0.3758	0.8926
2	0.3020	0.6778	0.6242
3	0.2013	0.8791	0.3222
4	0.0881	0.9672	0.1209
5	0.0264	0.9936	0.0328
6	0.0055	0.9991	0.0064
7	0.0008	0.9999	0.0009
8	0.0001	1.0000	0.0001
9	0.0000	1.0000	0.0000
10	0.0000	1.0000	0.0000

Figure 3.3.1. Spreadsheet template for Binomial Distribution
[Binomial in Distributions]

In cells C4 and D4, the values of n and p are entered. The probabilities of exactly x successes, at most x successes and at least x successes are tabulated in columns C, D and E. By changing the values of n and p suitably, many kinds of probabilities can be directly read off of the sheet.

Box 3.3.1. Formulas for Binomial Random Variable

If $X \sim B(n, p)$, then

$$P(X = x) = \binom{n}{x} p^x (1-p)^{(n-x)} \qquad \text{for } x = 0, 1, 2, \ldots, n$$

$E[X] = np$

$V(X) = np(1-p)$

Example: If $n = 5$; $p = 0.6$; then

$P(X=3) = 10 * 0.6^3 * 0.4^2 = 0.3456$

```
=BINOMDIST(3,5,0.6,FALSE)
```

$E[X] = 5 * 0.6 = 3$

$V(X) = 5 * 0.6 * 0.4 = 1.2.$

A nice thing about standard random variables, including the binomial, is that simple formulas are available for calculating their expected value, variance and other statistics. For the binomial random variable, the expected value is given by the formula $E[X] = np$ and the variance by $V(X) = np(1-p)$. These are summarized in Box 3.3.1.

3.3.2 Problem Solving

Suppose an operator wants to produce *at least* 2 good pins. [In practice, one would want *at least* some number of *good* things, or *at most* some number of *bad* things. Rarely would one want *exactly* some number of good or bad things.] He produces the pins using a lathe that has 0.6 probability of making a good pin in each trial, and this probability stays constant throughout. Suppose he produces 5 pins. What is the probability that he would have made at least 2 good ones? One way to answer the question would be to calculate $P(X = 2) + P(X = 3) + P(X = 4) + P(X = 5)$. Or, one may use the BINOMDIST function to get the *cumulative* probability $P(X \leq 1)$, and then calculate the answer as its complement, namely, $1-P(X \leq 1)$.

An easier way is to use the template shown in Figure 3.3.1. After making sure that n is filled in as 5 and p as 0.6, the answer is read off as 0.9130. That is, the operator can be 91.3% confident that he would have at least 2 good pins.

Let us go further with the problem. Suppose it is important that the operator has at least 2 good pins, and therefore he wants to be at least 99% confident that he would have at least 2 good pins. [In this type of situations, the phrases "at least" and "at most" occur often. You should read carefully.] With 5 trials, we just found that he can be only 91.3% confident. To increase the confidence level, he needs to make more trials. How many more? Using the spreadsheet template, we can answer this question by progressively increasing the value of n and stopping when P(At least 2) just exceeds 99%. On doing this, we find that 8 trials will do and 7 will not. Hence the operator should make at least 8 trials.

Now suppose the operator has enough time to produce only 5 pins, but still wants to have at least 99% confidence of producing at least 2 good pins, by improving the lathe and thus improving p. What should be the minimum value of p? Here again, we can keep increasing p and stop when P(At least 2) just exceeds 99%. This process could get tiresome if we need, say, four decimal place accuracy. This is where the **Goal seek...** command (see Chapter 0) in the spreadsheet comes in handy. Using the Goal seek facility, the answer is found to be 0.7777.

3.4 The Negative Binomial Distribution

3.4.1 Introduction

Consider again the case of the operator who wants to produce 2 good pins using a lathe that has 0.6 probability of making a good one in each trial. Under binomial distribution, we assumed that he produces 5 pins and calculated the probability of getting at least 2 good ones. In practice though, the operator would produce the pins one by one and stop when he gets 2 good ones. For instance, if the first two are good, then he would stop right there. If the first and the third are good, then he would stop with the third, and so on. Notice that in this scenario, the number of successes is held constant at 2, and the number of trials is random. The number of trials could be 2, 3, 4, ... and so on. [Contrast this with the binomial distribution where the number of trials is fixed and the number of successes is random.]

The number of trials made in this scenario is said to follow a *Negative Binomial Distribution.* Let *s* denote the exact number of successes desired and *p* the probability of success in each trial. Let *X* denote the number of trials made until the desired number of successes are achieved. Then *X* will follow a negative binomial distribution and we shall write $X \sim \text{NB}(s, p)$ where NB stands for Negative Binomial.

3.4.2 Negative Binomial Distribution Formulas

What is the formula for $P(X = x)$ when $X \sim \text{NB}(s, p)$? We note that the very last trial must be a success; otherwise, we would have already had the desired number of successes with $x-1$ trials, and we should have stopped right there. The last trial being a success, the first $x-1$ trials should have had $s-1$ successes. Thus the formula should be

$$P(X = x) = \binom{x-1}{s-1} p^s (1-p)^{(x-s)} \qquad (3.4.1)$$

In Excel, the function NEGBINOMDIST can be used to calculate probabilities. One must note that the first argument of this function is the number of *failures*, namely, $x - s$. The formula for $P(X = x)$ is:.

$P(X = x)$ = `NEGBINOMDIST(x-s,s,p)` (3.4.2)

The formula for the mean can be arrived at intuitively, For instance, if $p = 0.3$, and 3 successes are desired, then the expected number of trials to achieve 3 successes is 10. Thus the mean should have the formula $\mu = s/p$. The variance is given by the formula $\sigma^2 = s(1-p)/p^2$. These are presented in Box 3.4.1.

Box 3.4.1. Negative Binomial Distribution

If $X \sim \text{NB}(s, p)$, then

$$P(X = x) = \binom{x-1}{s-1} p^s (1-p)^{(x-s)} \qquad \text{for } x = s, s+1, s+2, \ldots$$

$E[X] = s/p$

$V(X) = s(1-p)/p^2$

Example: If $s = 2$; $p = 0.6$; then

$$P(X=5) = \binom{4}{1} * 0.6^2 * 0.4^3 = 0.0922$$

`=NEGBINOMDIST(3,2,0.6)`

$E[X] = 2 / 0.6 = 3.3333$

$V(X) = 2 * 0.4 / 0.6^2 = 2.2222.$

3.4.3 Problem Solving

In the template Distributions, there is a sheet for negative binomial distribution which can be used for solving problems. This template is reproduced in Figure 3.4.1 below.

Negative Binomial Distribution

s	p	
3	0.7274	

s = Number of successes desired
p = Probability of success in each trial
x = Number of trials needed

Mean	Variance	Stdev.
4.12455	1.54609	1.24342

x	P(Exactly x)	P(At most x)	P(At least x)
3	0.3848	0.3848	1.0000
4	0.3147	0.6995	0.6152
5	0.1716	0.8712	0.3005
6	0.0780	0.9492	0.1288
7	0.0319	0.9811	0.0508
8	0.0122	0.9932	0.0189
9	0.0044	0.9977	0.0068
10	0.0016	0.9992	0.0023
11	0.0005	0.9997	0.0008
12	0.0002	0.9999	0.0003
13	0.0001	1.0000	0.0001
14	0.0000	1.0000	0.0000
15	0.0000	1.0000	0.0000

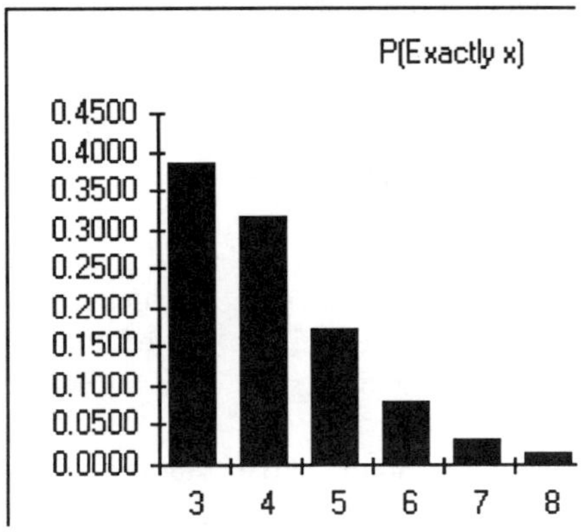

Figure 3.4.1. Negative Binomial Distribution Template
[Negative Binomial in Distributions]

The formulas in B19:E19 can be copied down further if more rows of results are desired. Note that in the range C10:C19, Formula 3.4.2 has been employed.

Let us return back to the operator who wants to keep producing pins until he has 2 good ones. The probability of getting a good one at any trial is 0.6. What is the probability that he would produce exactly 5? Looking at the template in Figure 3.4.1, we see that the answer is 0.0922, which agrees with the calculation in Box 3.4.1. We can, in addition, see that the probability of producing at most 5 is 0.9130 and at least 5 is 0.1792.

Suppose the operator has enough time to produce only 4 pins. How confident can he be that he would have 2 good ones within the available time? Looking at the template, we see that the probability of needing at most 4 trials is 0.8208 and hence he can be about 82% confident.

If he wants to be at least 95% confident, at least how many trials should he be prepared for? Looking at the template, we infer that he should be prepared for at least 6 pins, since 5 trials yield only 91.30% confidence and 6 trials yield 95.90%.

Suppose the operator has enough time to produce only 4 pins and still wants to have at least 95% confident of getting 2 good pins within the available time. Suppose further, he wants to achieve this by increasing the value of *p*. What is the minimum *p* that would achieve this? Using the **Goal Seek** command, this can be answered as 0.7514. Specifically, you set cell D7 to 0.95 by changing cell C3.

3.5 The Geometric Distribution

3.5.1 Introduction

In a negative binomial distribution, the number of desired successes *s* can be any number. But in some practical situations, the number of successes desired is just 1. For instance, if one is attempting to pass a test, or get some information, it is enough to have just one success. Let *X* be the (random) number of Bernoulli trials, each having *p* probability of success, required to achieve just one success. *X* is then said to follow a *Geometric Distribution*, and we shall write $X \sim G(p)$. Note that the geometric distribution is a special case of negative binomial distribution where $s = 1$. The reason for the name "geometric distribution" is that the sequence of probabilities $P(X = 1)$, $P(X = 2)$, ... follows a *geometric progression*.

3.5.2 Geometric Distribution Formulas

Because geometric distribution is a special case of negative binomial distribution where s = 1, the formulas for negative binomial distribution with s fixed as 1 can be used for geometric distribution. Indeed, there is no Excel function for directly calculating geometric distribution probabilities, and the negative binomial formula can be used as shown in Box 3.5.1 below. Recall that the NEGBINOMDIST function works with number of failures rather than the total number of trials.

Box 3.5.1. Geometric Distribution Formulas

If $X \sim G(p)$, then

$P(X = x) = p(1-p)^{(x-1)}$ for $x = 1, 2, \ldots$

$E[X] = 1/p$

$V(X) = (1-p)/p^2$

Example: If $p = 0.6$, then

$P(X=5) = 0.6 * 0.4^4 = 0.0154$

```
=NEGBINOMDIST(4,1,0.6)
```

$E[X] = 1 / 0.6 = 1.6667$

$V(X) = 0.4 / 0.6^2 = 1.1111.$

3.5.3 Problem Solving

Consider the operator who produces pins one by one on a lathe that has 0.6 probability of producing a good one at each trial. Suppose he wants only one good pin and stops as soon as he gets one. What is the probability that he would produce exactly 5 pins? In the Distributions workbook there is a sheet named Geometric. On that sheet, one enters the value 0.6 for p. See Figure 3.5.1. The answer can now be read off as 0.0154, same as the example calculation in Box 3.5.1. One can further read on the spreadsheet that the probability of at most 5 is 0.9898 and at least 5 is 0.0256. Also note that the probability of exactly 1, 2, 3, ... trials follows the sequence 0.6, 0.24, 0.096, 0.0384, ... which is indeed a geometric progression.

Geometric Distribution

p
0.6

Mean	Variance	Stdev.
1.66667	1.11111	1.05409

x	P(Exactly x)	P(At most x)	P(At least x)
1	0.6000	0.6000	1.0000
2	0.2400	0.8400	0.4000
3	0.0960	0.9360	0.1600
4	0.0384	0.9744	0.0640
5	0.0154	0.9898	0.0256
6	0.0061	0.9959	0.0102
7	0.0025	0.9984	0.0041
8	0.0010	0.9993	0.0016
9	0.0004	0.9997	0.0007
10	0.0002	0.9999	0.0003
11	0.0001	1.0000	0.0001
12	0.0000	1.0000	0.0000
13	0.0000	1.0000	0.0000
14	0.0000	1.0000	0.0000

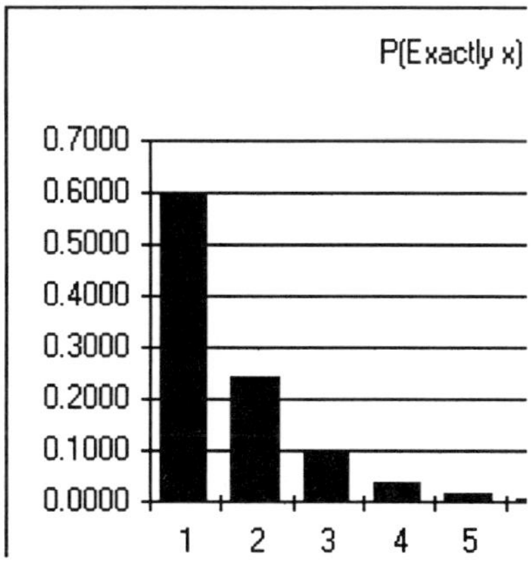

Figure 3.5.1 Geometric Distribution Template
[Geometric in Distributions]

Now suppose the operator has time enough for at most 2 pins, how confident can he be of getting a good one within the available time? From the template, the answer is 0.8400, or 84%. What if he want to be at least 95% confident? Again from the template, he must have enough time for 4 pins, because 3 would yield only 93.6% confidence and 4 yields 97.44%.

3.6 The Hypergeometric Distribution

3.6.1 Introduction

Assume that a box contains 10 pins of which 6 are good and the rest defective. An operator picks 5 pins at random from the 10, and is interested in the number of good pins picked. Let X denote the number. We should first note that X is not a binomial random variable, because the probability of success p, which is the probability of picking a good pin is neither constant nor independent from trial to trial. The first pin picked has 0.6 probability of being good; the second has either 5/9 or 6/9 probability depending on whether or not the first was good. Therefore, X does not follow binomial distribution, but follows what is called a *Hypergeometric Distribution.* In general, when a pool of size N contain S successes and $(N - S)$ failures and a random sample of size n is drawn from the pool, the number of successes X in the sample follows a hypergeometric distribution. We shall then write $X \sim HG(n, S, N)$. The situation is depicted in Figure 3.6.1.

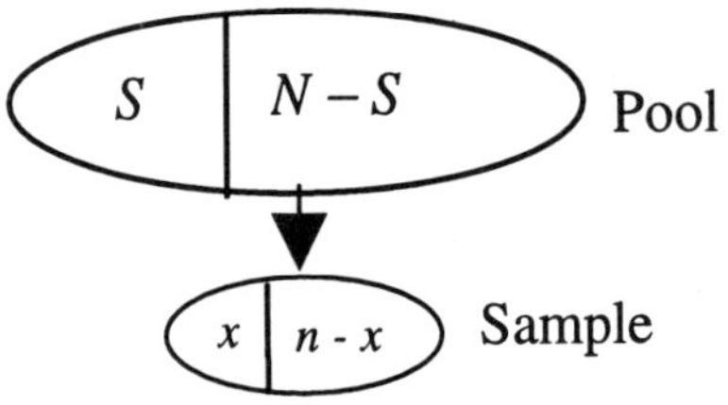

Figure 3.6.1 Where Hypergeometric Distribution Occurs

3.6.2 Hypergeometric Distribution Formulas

Let us derive the formula for $P(X = x)$ when X is hypergeometrically distributed. The x number of successes have to come from the S successes in the pool which can happen in $\binom{S}{x}$ ways. The $(n - x)$ failures have to come from the $(N - S)$ failures in the pool which can happen in $\binom{N-S}{n-x}$ ways. Together, the x successes and $(n - x)$ failures can happen in $\binom{S}{x}\binom{N-S}{n-x}$ ways. Finally, there are $\binom{N}{n}$ ways of selecting a sample of size n. Putting them all together,

$$P(X = x) = \frac{\binom{S}{x}\binom{N-S}{n-x}}{\binom{N}{n}} \qquad (3.6.1)$$

In this formula n cannot exceed N since the sample size cannot exceed the pool size. The minimum possible value for x is $\text{Max}(0, n - N + S)$ and the maximum possible value is $\text{Min}(n, S)$. For instance, if $n = 9$, $S = 5$, and $N = 12$, you may verify that there would be at least 2 successes and at most 5.

Box 3.6.1 Hypergeometric Distribution Formulas

If $X \sim \text{HG}(n, S, N)$, then

$$P(X = x) = \frac{\binom{S}{x}\binom{N-S}{n-x}}{\binom{N}{n}} \quad \text{for } Max(0,n\text{-}N\text{+}S) < x < Min(n,S)$$

$E[X] = np$ where $p = S/N$

$$V(X) = np(1-p)\left[\frac{N-n}{N-1}\right]$$

Example: If $n = 5$, $S = 6$, $N = 10$, then

$$P(X = 2) = \frac{\binom{6}{2}\binom{10-5}{5-1}}{\binom{10}{5}} = 0.2381$$

`=HYPGEOMDIST(2,5,6,10)`

$E[X] = 5 * (6/10) = 3.00$

$V(X) = 5*0.6*(1–0.6)*(10–5)/(10–1) = 0.6667$

The proportion of successes in the pool, which is the ratio S/N, is the probability of the first trial being a success. This ratio is denoted by the symbol p since it resembles the p used in binomial distribution. The expected value and variance of X are expressed using p as:

$$E[X] = np \tag{3.6.2}$$

$$V(X) = np(1-p)\left[\frac{N-n}{N-1}\right]. \tag{3.6.3}$$

Notice that the formula for $E[X]$ is the same as for the binomial case. The formula for the $V(X)$ is similar to but not the same as the binomial case. The difference is the additional factor in square brackets. This factor approaches 1 as N becomes larger and larger compared to n and may be dropped for large N.

In Excel, hypergeometric probabilities are calculated using the HYPGEOMDIST function. For given values of x, n, S and N,

$P(X = x) =$ `HYPGEOMDIST(x,n,S,N)` (3.6.4)

All of the above formulas are summarized in Box 3.6.1.

3.6.3 Problem Solving

The sheet named Hypergeo in the Distributions workbook is shown in Figure 3.6.2. The input cells here are for *n, S, N*. Since Excel does not distinguish between upper and lower case letters, *n* and *N* cannot be used together. Therefore *N* has been labeled *NN*. In the range C7:C20, Formula 3.6.4 has been used.

Let us consider the case where a box contains 10 pins out of which 6 are good, and the operator picks 5 at random. What is the probability that exactly 2 good pins are picked. The answer 0.2381 is in cell C8. Additionally, the probability that at most 2 and at least two good ones are picked are, respectively, 0.2619 and 0.9762.

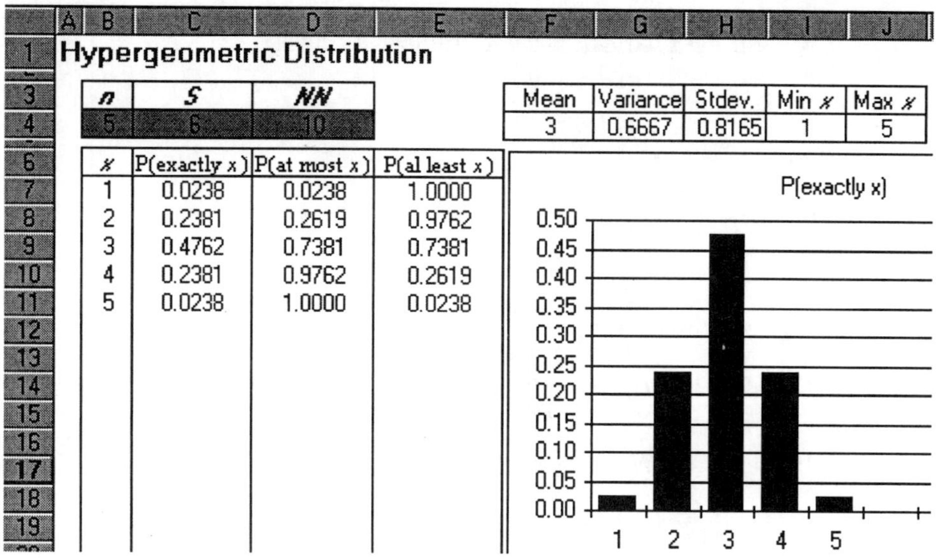

Figure 3.6.2 Hypergeometric Distribution Template
[Hypergeo in Distributions]

3.7 The Poisson Distribution

3.7.1 Introduction

Consider a lathe that produces pins in mass and rarely produces a defective one. Let us say it produces 20,000 pins and has 1/10,000 chance of producing a defective one. Suppose we are interested in the number of defective pins produced. One may try to do this by using the binomial distribution with n = 20000 and p = 1/10000. But the calculation would be tedious because n is a very large number, p is a very small number and the binomial formula calls for $n!$ and p^{n-x} which are hard to calculate even on a computer. However, the expected number of defectives produced is 20000 * (1/10000) = 2, which is neither too large nor too small. It turns out that when n is very large, if the expected value $\mu = np$ is neither too large nor too small, say, lies between 0.01 and 30, then the binomial formula for P(X = x) can be approximated as

$$P(X = x) = \frac{e^{-\mu}\mu^{x}}{x!} \qquad \text{for } x = 0, 1, 2, \ldots \qquad (3.7.1)$$

This formula is known as the *Poisson* formula, and the distribution is called the *Poisson Distribution.* In general, if the occurrence of an event is rare, and we count the number of such events occurring during a fixed interval, then that number would follow a Poisson distribution. If X follows a Poisson distribution, we shall write $X \sim P(\mu)$ where μ is the expected value of the distribution. Note that just one number μ is enough to describe the distribution, and in this sense the Poisson distribution is a simple one, even simpler than binomial. Coming to the variance of Poisson distribution, we note that the binomial variance is $np(1-p)$. But since p is very small, $(1-p)$ is close to 1 and therefore can be omitted. Thus the Poisson variance is np which is the same as the mean.

$$V(X) = np = \mu. \qquad (3.7.2)$$

These formulas are tabulated in Box 3.7.1 below.

Box 3.7.1 Poisson Distribution Formulas

If $X \sim P(\mu)$, then

$$P(X = x) = \frac{e^{-\mu}\mu^{x}}{x!} \qquad \text{for } x = 0, 1, 2, \ldots$$

$$E[X] = np = \mu$$
$$V(X) = np = \mu$$

Example: If $\mu = 2$, then

$$P(X = 3) = \frac{e^{-2}2^{3}}{3!} = 0.1804$$

```
=POISSON(3,2,FALSE)
```

$E[X] = \mu = 2.00$

$V(X) = \mu = 2.00$

In Excel, $P(X = x)$ can be calculated using the formula

```
=POISSON(x,m,FALSE)
```
(3.7.3)

The last argument, FALSE, is to specify that we want the probability of X being exactly equal to x. If it is specified as TRUE, then the *cumulative* probability of X being at most x would be returned. The Poisson sheet from Chapter3.xlt is shown in Figure 3.7.1 below.

Poisson Distribution

Mean	Variance	Stdev
4	4	2

x	P(Exactly x)	P(At most x)	P(At least x)
0	0.0183	0.0183	1.0000
1	0.0733	0.0916	0.9817
2	0.1465	0.2381	0.9084
3	0.1954	0.4335	0.7619
4	0.1954	0.6288	0.5665
5	0.1563	0.7851	0.3712
6	0.1042	0.8893	0.2149
7	0.0595	0.9489	0.1107
8	0.0298	0.9786	0.0511
9	0.0132	0.9919	0.0214
10	0.0053	0.9972	0.0081
11	0.0019	0.9991	0.0028
12	0.0006	0.9997	0.0009
13	0.0002	0.9999	0.0003

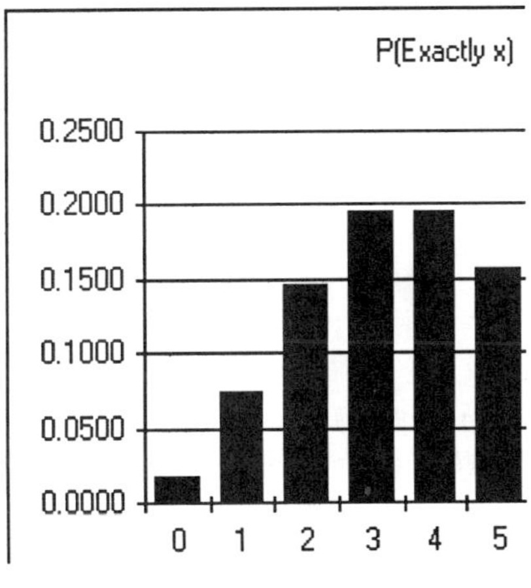

Figure 3.7.1 Poisson Distribution Template

The only input cell C4 is for the expected value μ. One can read from the template the fact that when $\mu = 2$, the probability of exactly three successes is 0.1804, at most three successes is 0.8571 and at least three successes is 0.3233.

3.8 Continuous Random Variables

When a random variable is continuous, we cannot use a table of x and $P(x)$ values, as in Figure 3.1.1, to describe the distribution because there will be uncountably many x values. Instead, we use a *probability density function* $f(x)$ to describe the distribution. If $f(x)$ is the probability density function of X, then the probability that X will be between x_1 and x_2 is given by the formula

$$P(x_1 \le X \le x_2) = \int_{x_1}^{x_2} f(x)dx$$

which, pictorially, is the area below the curve $f(x)$ between x_1 and x_2. Figure 3.8.1 shows this area.

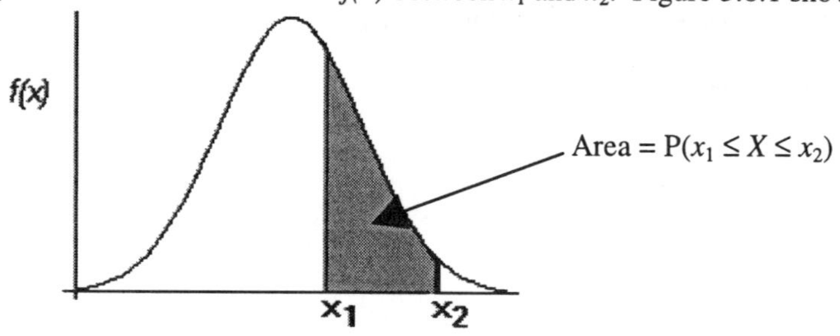

Figure 3.8.1 Probability Density Function $f(x)$

Some restrictions on $f(x)$ are natural. Because a probability cannot be negative, $f(x)$ cannot be negative over any interval; and because the total probability should be 1, the area under $f(x)$ should be 1.

As in the case of discrete random variables continuous random variables that occur in practice can also be approximated to one of the standard ones if we know what type of random variable it is. We shall see two standard continuous distributions. In the next chapter we shall see the third, namely, the normal distribution.

3.9 Uniform Distribution

The uniform distribution is the simplest of continuous distributions. The probability density function is

$$f(x) = 1/(b - a) \quad \text{for } a \le x \le b$$
$$= 0 \quad \text{for all other } x.$$

where a is the minimum possible value and b is the maximum possible value of X. The graph of $f(x)$ is shown in Figure 3.9.1.

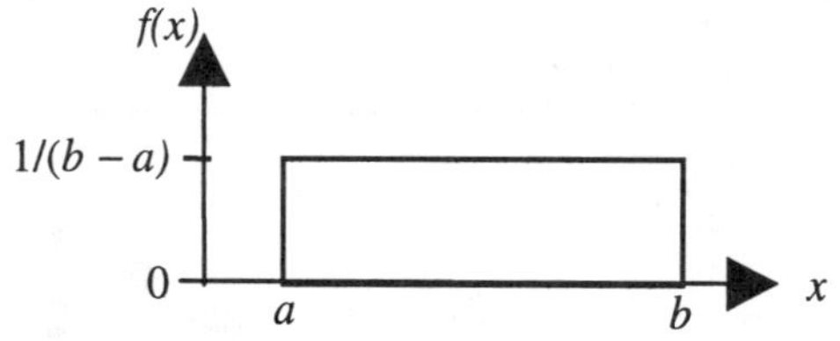

Figure 3.9.1 Uniform Distribution

Because the curve of $f(x)$ is a flat line, the area under it between any two points x_1 and x_2 will be a rectangle with height $1/(b - a)$ and width $(x_2 - x_1)$. Thus $P(x_1 \le X \le x_2) = (x_2 - x_1)/(b - a)$. If X is uniformly distributed between a and b we shall write $X \sim U(a, b)$.

The mean of the distribution is the midpoint between a and b which is $(a + b)/2$. Using integration, it can be shown that the variance is $(b - a)^2/12$. The skewness is zero (why?). The distribution is flat, and has a relative kurtosis of -1.2. The formulas for uniform distribution are summarized in Box 3.9.1. Because the probability calculation is simple, there is no special spreadsheet function for uniform distribution. Box 3.9.1 also contains some example calculations.

A common instance of uniform distribution is waiting time for a facility that goes in cycles. A shuttle bus or an elevator are good examples. They move, roughly, in cycles with some cycle time. If the user comes to a stop at a random time and waits till the facility arrives, the waiting time will be uniformly distributed between a minimum of zero and a maximum equal to the cycle time. In other words, if a shuttle bus has a cycle time of 20 minutes, the waiting time would be uniformly distributed between 0 and 20 minutes.

Occasionally, customer service times, such as the time taken by a barber to give a haircut, are found to be uniformly distributed.

Box 3.9.1 Uniform Distribution Formulas

If $X \sim \text{U}(a, b)$, then

$$f(x) = 1/(b-a) \qquad \text{for } a \le x \le b$$
$$= 0 \qquad \text{for all other } x.$$
$$\text{P}(x_1 \le X \le x_2) = (x_2 - x_1)/(b-a) \qquad \text{for } a \le x_1 < x_2 \le b$$
$$\text{E}[X] = (a+b)/2$$
$$\text{V}(X) = (b-a)^2/12$$

Example: If $a = 10$ and $b = 20$, then

$$\text{P}(12 \le X \le 18) = (18-12)/(20-10) = 0.6$$
$$\text{E}[X] = (10+20)/2 = 15$$
$$\text{V}(X) = (20-10)^2/12 = 8.3333$$

To solve problems, it is convenient to see many probabilities at the same time, for which a spreadsheet template works well. Figure 3.9.2 shows the template in the Distributions workbook. If $X \sim$ U(10, 20), what is $\text{P}(12 \le X \le 18)$? In the template, make sure the Min and Max are set to 10 and 20 in cells B4 and C4. Enter 12 and 18 in cells E7 and F7. The answer of 0.6 appears in cell G7. If probability of at least or at most x is needed, columns A through C may be used. Inverse calculations are possible in columns I through M, where for a given probability of at least or at most x, the corresponding x value is calculated. As usual, one may also use facilities such as the **Goal Seek** command or the **Solver** tool in conjunction with this template.

	A	B	C	E	F	G	I	J	L	M
1	Uniform Distribution									
3		Min	Max	Var.	Stdev.					
4		10	20	8.3333	2.8868					
5							Inverse calculations			
6	x	P(<=x)	P(>=x)	x1	x2	P(x1<X<x2)	P(<=x)	x	P(>=x)	x
7	12	0.2000	0.8000	12	18	0.6000	0.2	12	[illegible]	17
8				11	12	0.1000	[illegible]			
9	22	1.0000	0.0000	11	22	0.9000			0	20
10				21	22	0.0000	0	10		
11									1	10

Figure 3.9.2 Template for Uniform Distribution
[Uniform in Distribution]

3.10 The Exponential Distribution

Suppose an event occurs with an average frequency of λ occurrences per hour and this average frequency is constant in that the probability that the event will occur during any tiny duration t is λt. Suppose further we arrive at the scene at any given time and wait till the event occurs. The waiting time will then follow an *exponential distribution.* This distribution is the continuous limit of the geometric distribution. Suppose our waiting time was x. For the event (or success) to occur at time x, every tiny duration t from time 0 to time x should be a failure and the interval x to $x+t$ must be a success. This is nothing but a geometric distribution. To get the continuous version, we take the limit of this process at t approaches zero.

Exponential distribution is fairly common in practice. The time between two successive break downs of a machine will be exponentially distributed, and is relevant to maintenance engineers. The mean μ in this case is known as the *Mean Time Between Failures* or MTBF. The life of a product that fails by accident rather than by wear and tear also follows exponential distribution and is relevant to warranty policies. When X is exponentially distributed with frequency λ, we shall write $X \sim E(\lambda)$.

The probability density function $f(x)$ of exponential distribution has the form

$$f(x) = \lambda\, e^{-\lambda x}$$

where λ is the frequency with which the event occurs. The frequency λ is expressed as so many times per unit time, such as 1.2 times per month. The mean of the distribution is $1/\lambda$ and the variance is $(1/\lambda)^2$. Just as the geometric distribution, the exponential distribution is positively skewed.

The formula

```
=EXPONDIST(x,λ,TRUE)
```

will return the cumulative probability $P(X \le x)$ and the formula

```
=EXPONDIST(x,λ,FALSE)
```

will return the value of $f(x)$ at x. These formulas are used in the template in Chapter3.xlt. The template is seen in Figure 3.10.1. Box 3.10.1 summarizes the formulas and provides example calculations.

Box 3.10.1 Exponential Distribution Formulas

If $X \sim E(\lambda)$, then

$f(x) = \lambda\, e^{-\lambda x}$ for $x \ge 0$

```
=EXPONDIST(x,λ,FALSE)
```

$P(X \le x) = 1 - e^{-\lambda x}$ for $x \ge 0$

```
=EXPONDIST(x,λ,TRUE)
```

$P(X \ge x) = e^{-\lambda x}$ for $x \ge 0$

$P(x_1 \le X \le x_2) = e^{-x_1} - e^{-x_2}$ for $0 \le x_1 < x_2$

$E[X] = 1/\lambda$

$V(X) = 1/\lambda^2$

Example: If $\lambda = 1.2$, then

$P(X \le 0.5) = 1 - \varepsilon^{-1.2*0.5} = 0.4512$

```
=EXPONDIST(0.5,1.2,TRUE)
```

$P(X \ge 0.5) = \varepsilon^{-1.2*0.5} = 0.5488$

$P(1 \le X \le 2) = e^{-1} - e^{-2} = 0.2105$

$E[X] = 1/1.2 = 0.8333$

$V(X) = 1/1.2^2 = 0.6944$

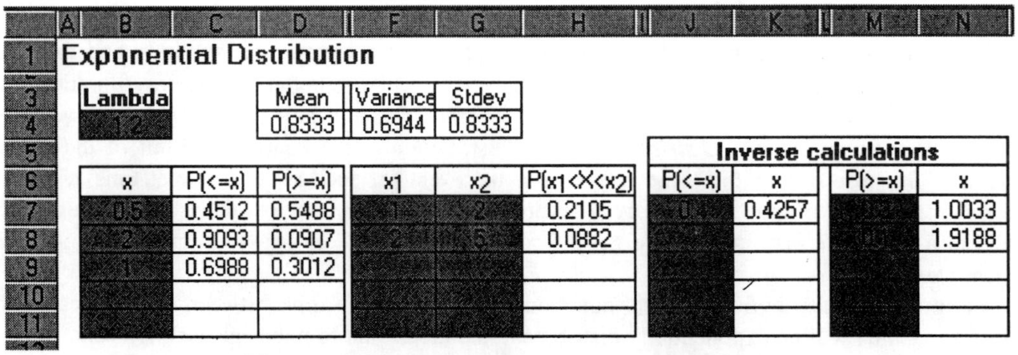

Exponential Distribution

Lambda		Mean	Variance	Stdev
1.2		0.8333	0.6944	0.8333

x	P(<=x)	P(>=x)	x1	x2	P(x1<X<x2)	P(<=x)	x	P(>=x)	x
						Inverse calculations			
0.5	0.4512	0.5488	1	2	0.2105	0.4	0.4257	[illegible]	1.0033
2	0.9093	0.0907	2	[illegible]	0.0882			[illegible]	1.9188
1	0.6988	0.3012							

Figure 3.10.1 Exponential Distribution Template
[Exponential in Distributions]

In order to use the exponential distribution template shown in Figure 3.10.1, the value of λ must be entered in cell B4. At times, the mean μ rather than λ may be known, and its reciprocal $1/\mu = \lambda$ is what should be entered in cell B4. The green cells are input cells and the rest are protected. As usual, the **Goal Seek** command and the **Solver** tool can be used in conjunction with this template to solve problems.

3.11 Summary

In this chapter we have seen several important standard random variables, the associated formulas and problem solving with spreadsheets. In order to use a spreadsheet template, one should know *which* template to use, and to know that one should know what kind of random variable one has at hand. In this summary, we shall concentrate on this question.

A discrete random variable X will follow a binomial distribution if it is the number of successes in n independent Bernoulli trials. Make sure that the probability of success, p, remains constant in all trials. X will follow a negative binomial distribution if it is the number of Bernoulli trials made to achieve a desired number of successes. It will follow a geometric distribution when the desired number of successes is 1. X will follow a hypergeometric distribution if it is the number of successes in a random sample drawn from a finite pool of successes and failures. X will follow a Poisson distribution if it is the number of occurrences of a rare event during a finite period.

Waiting time for an event that occurs periodically is uniformly distributed. Waiting time for a rare event is exponentially distributed.

3.12 Exercises

1. A random variable X is described in each of the following items. Find out which standard distribution X is likely to follow in each case.

i. X is the number of calls received at a 911 telephone number on a given day.

ii. Your friend is equally likely to be found at any one of a number of places during coffee break. X is the number of places you have to look before you find him.

iii. A drawer contains 8 red socks and 12 green socks. You pick 6 socks at random from the drawer. X is the number of red socks you picked.

iv. An assembly line breaks down and stops about once in six months. X is the number of times it breaks down in a given year.

v. X is the time gap between two successive crashes of a mainframe computer.

vi. A company buys bolts in bulk and rarely finds defective ones. X is the number of defective bolts in a shipment that has just arrived.

vii. Twenty students take a test in which each of them has 30% chance of getting an A grade. X is the number of students who get A grades.

viii. You need 6 volunteers for a task and you ask your friends, each of whom has 60% chance of agreeing, one by one until you get 6 volunteers. X is the number of friends you have to ask.

ix. Fifty guests are sent invitations for a party, and each of them has 80% chance of attending the party. X is the number of guests who actually show up.

x. A factory has a number of machines that occasionally break down. X is the number of machines that break down on a given day.

xi. A machine breaks down occasionally. X is the time gap between two successive breakdowns.

xii. Headline News on TV starts every half hour. You switch on the TV at a random time and wait till the start. X is the time you have to wait.

xiii. A committee is formed by picking members at random from the US Senate. X is the number of Democrats picked.

xiv. An airline has sold 132 tickets for a flight. Each ticket buyer has 96% chance of showing up. X is the actual number of passengers that show up.

xv. A company sends out tens of thousands of free samples to prospective customers each of whom has a very small chance of buying the product. X is the actual number of customers who buy the product.

xvi. A machine breaks down occasionally. You visit the machine and wait till it breaks down. X is the time you have to wait.

xvii. A shuttle bus leaves an airport every 45 minutes. You arrive at a random time and wait for it. X is the time you have to wait.

Binomial Distribution

2. A salesperson goes door-to-door in a residential area and demonstrates the use of a new household appliance. At the end of a demonstration, there is a constant 0.2107 probability that the potential customer would place an order for the product. To perform satisfactorily on the job, the salesperson needs at least 4 orders. Assume that each demonstration is a Bernoulli trial.

a. If the salesperson makes 15 demonstrations, what is the probability that there would be exactly 4 orders?

b. If the salesperson makes 16 demonstrations, what is the probability that there would be at most 4 orders?

c. If the salesperson makes 17 demonstrations, what is the probability that there would be at least 4 orders?

d. If the salesperson makes 18 demonstrations, what is the probability that there would be anywhere from 4 to 8 (both inclusive) orders?

e. If the salesperson wants to be at least 90% confident of getting at least 4 orders, at least how many demonstrations should she make?

f. The salesperson has time to make only 22 demonstrations, and still wants to be at least 90% confident of getting at least 4 orders. She intends to gain this confidence by improving the quality of her demonstration and thereby improving the chances of getting an order at the end of a demonstration (currently this probability is 0.2107). At least to what value should this probability be increased in order to gain the desired confidence? Your answer should be accurate to 4 decimal places.

3. An MBA graduate is applying for 9 jobs, and believes she has in each of the 9 cases a constant and independent 0.48 probability of getting an offer.

a. What is the probability that she will have at least 3 offers.

b. If she wants to be 95% confident of having at least 3 offers, how many more jobs should she apply for? (Assume each of these additional applications will also have the same probability of success.)

c. If there are no more than the original 9 jobs that she can apply to, what value of probability of success would give her 95% confidence of at least 3 offers?

4. A computer laboratory in a school has 33 computers. Each of the 33 computers has 90% reliability. Allowing for, roughly, 10% of the computers to be down, an instructor specifies an enrollment ceiling of 30 for his class. Assume that a class of 30 students is taken to into the lab.

a. What is the probability that each of the 30 students will get a computer in working condition?

b. The instructor is surprised to see the low value of the answer to the above question and decides to improve it to at least 95% by doing one of the following:

i) Decrease the enrollment ceiling.
ii) Increase the number of computers in the lab.
iii) Increase the reliability of all the computers.

To help the instructor, find out what the increase or decrease should be for each of the three alternatives.

5. A commercial jet aircraft has 4 engines. In order for an aircraft in flight to land safely, at least 2 engines should be in working condition. Each engine has an independent reliability of 92%.

a. What is the probability that an aircraft in flight can land safely?

b. If the answer to part a. above has to be at least 99.5%, what is the minimum value for p?

Negative Binomial Distribution

6. A graduating student keeps applying for jobs until she has 3 offers. The probability of getting an offer at any trial is 0.48.

a. What is the expected number of applications? What is the variance?

b. If she has enough time to complete only 6 applications, how confident can she be of getting 3 offers within the available time?

c. If she wants to be at least 95% confident of getting 3 offers, how many applications should she be prepared for?

d. Suppose she has time for at most 6 applications. For what minimum value of p can she still have 95% confidence of getting 3 offers within the available time?

7. A real estate agent has 4 houses to sell before the end of the month by contacting prospective customers one by one. Each customer has an independent 0.24 probability of buying a house on being contacted by the agent.

a. If the agent has enough time to contact only 15 customers, how confident can she be of selling all 4 houses within the available time?

b. If the agent wants to be at least 70% confident of selling all the houses within the available time, at least how many customers should she contact? [If necessary extend the template downward to more rows.]

c. What minimum value of p will yield 70% confidence of selling all 4 houses by contacting at most 15 customers?

d. In order to answer part c. above more thoroughly, tabulate the confidence for p values ranging from 0.2 to 0.6 in steps of 0.05.

Geometric Distribution

8. A graduating student keeps applying for jobs until she gets an offer. The probability of getting an offer at any trial is 0.35.

a. What is the expected number of applications? What is the variance?

b. If she has enough time to complete at most 4 applications, how confident can she be of getting an offer within the available time?

c. If she wants to be at least 95% confident of getting an offer, how many applications should she be prepared for?

d. Suppose she has time for at most 4 applications. For what minimum value of p can she have 95% confidence of getting an offer within the available time?

Hypergeometric Distribution

9. A shipment of pins contains 25 good ones and 2 defective ones. At the receiving department, an inspector picks 3 pins at random and tests them. If any defective pin is found among the three that are tested, the shipment would be rejected.

a. What is the probability that the shipment would be accepted?

b. In order to increase the probability of acceptance to at least 90%, it is decided to do one of the following:

i) add some good pins to the shipment
ii) remove some defective pins in the shipment

For each of the two options, find out exactly how many should be added or removed.

10. A committee of 7 members is to be formed by selecting members at random from a pool of 14 candidates consisting of 5 women and 9 men.

a. What is the probability that there will be at least 3 women in the committee?

b. It is desired to increase the chance that there are at least 3 women in the committee to 80% by doing one of the following.

i) adding more women to the committee
ii) removing some men from the committee

For each of the two options, find out how many should be added or removed.

Poisson Distribution

11. A mainframe computer in a university crashes on the average 0.71 times in a semester.

a. What is the probability that it will crash at least 2 times in a given semester?

b. What is the probability that it will not crash at all in a given semester?

c. It is desired to increase the probability of no crash at all in a semester to at least 90%. What is the largest μ that will achieve this goal?

12. The number of rescue calls received by a rescue squad in a city follows a Poisson distribution with μ = 2.83 per day. The squad can handle at most 4 calls a day.

a. What is the probability that the squad will be able to handle all the calls on a particular day?

b. The squad wants to have at least 95% confidence of being able to handle all the calls received in a day. At least how many calls a day should the squad be prepared for?

c. Assuming that the squad can handle at most 4 calls a day, what is the largest value of μ that would yield 95% confidence that the squad can handle all calls?

Uniform Distribution

13. A student takes the campus shuttle bus to reach the classroom building. The shuttle bus arrives at his stop every 15 minutes but the actual arrival time at the stop is random. The student allows 10 minutes waiting time for the shuttle in his plan to make it in time to the class.

a. What is the expected waiting time? What is the variance?

b. What is the probability that the wait will be between 4 and 6 minutes?

a. What is the probability that the student will be in time for the class?

b. If he wants to be 95% confident of making it in time for the class, how much time should he allow for waiting for the shuttle?

Exponential Distribution

14. A hydraulic press breaks down at the rate of 0.1742 times per day.

a. What is the MTBF?

b. On a given day, what is the probability that it will break down?

c. If four days have passed without a break down, what is the probability that it will break down on the fifth day?

d. What is the probability that five consecutive days will pass without any break down?

15. Laptop computers produced by a company have average life of 38.36 months. Assume that the life of a computer is exponentially distributed (which is a good assumption).

a. What is the probability that a computer will fail within 12 months?

b. If the company gives a warranty period of 12 months, what proportion of computers will fail during the warranty period?

c. Based on the answer to part b., would you say the company can afford to give a warranty period of 12 months?

d. If the company wants not more than 5% of the computers to fail during the warranty period, what should be the warranty period?

e. If the company wants to give a warranty period of 3 months and still wants not more than 5% of the computers to fail during the warranty period, what should be the minimum average life of the computers?

3.13 Projects

1. In most statistics textbooks, you will find cumulative binomial probability tables in the format shown below. These can be created using spreadsheets using the BINOMDIST and DATA/TABLE commands.

n = 5

		p								
		0.1	0.2	0.3	0.4	0.5	0.6	0.7	0.8	0.9
x	0	0.5905	0.3277	0.1681	0.0778	0.0313	0.0102	0.0024	0.0003	0.0000
	1	0.9185	0.7373	0.5282	0.3370	0.1875	0.0870	0.0308	0.0067	0.0005
	2	0.9914	0.9421	0.8369	0.6826	0.5000	0.3174	0.1631	0.0579	0.0086
	3	0.9995	0.9933	0.9692	0.9130	0.8125	0.6630	0.4718	0.2627	0.0815
	4	1.0000	0.9997	0.9976	0.9898	0.9688	0.9222	0.8319	0.6723	0.4095

a. Create the above table.
b. Create a similar table for n = 7.

2. Look at the shape of the binomial distribution for various combinations of n and p. Specifically, let $n = 5$ and try $p = 0.2$, 0.5 and 0.8. Repeat the same for other values of n. Can you say something about how the skewness of the distribution is affected by p and n?

3. What can you say about the skewness of the negative binomial distribution? Try various values of s and p on the template.

4. An MBA graduate keeps interviewing for jobs, one by one, and will stop interviewing on receiving an offer. In each interview he has an independent probability 0.2166 of getting the job.

a. What is the expected number of interviews? What is the variance?

b. If there is enough time for only 6 interviews, how confident can he be of getting a job within the available time?

c. If he wants to be at least 95% confident of getting a job, how many interviews should he be prepared for?

d. Suppose there is enough time for at most 6 interviews. For what minimum value of p can he have 95% confidence of getting a job within the available time?

e. In order to answer part d. above more thoroughly, tabulate the confidence level for p values ranging from 0.1 to 0.5 in steps of 0.05.

5. A shipment of thousands of pins contains some percentage of defectives. To decide whether or not to accept the shipment, the consumer follows a *sampling plan* where 80 items are chosen at random from the sample and tested. If the number of defectives in the sample is at most 3, the shipment is accepted. [The number 3 is known as the *acceptance number* of the sampling plan.]

a. Assuming that there are 3% defectives in the shipment, what is the probability that the shipment will be accepted? [*Hint: Use binomial distribution.*]

b. Assuming that there are 6% defectives in the shipment, what is the probability that the shipment will be accepted?

c. Using the **DATAITABLE** command, tabulate the probability of acceptance for defective percentage ranging from 0% to 15% in steps of 1%.

d. Plot a line graph of the table created in part c. above. [This graph is known as the *Operating Characteristic Curve* of the sampling plan.]

6. A shipment of 100 pins contains some defectives. To decide whether or not to accept the shipment, the consumer follows a *sampling plan* where 15 items are chosen at random from the sample and tested. If the number of defectives in the sample is at most 1, the shipment is accepted. [The number 1 is known as the *acceptance number* of the sampling plan.]

a. Assuming that there are 5% defectives in the shipment, what is the probability that the shipment will be accepted? [*Hint: Use hypergeometric distribution.*]

b. Assuming that there are 8% defectives in the shipment, what is the probability that the shipment will be accepted?

c. Using the **DATAITABLE** command, tabulate the probability of acceptance for defective percentage ranging from 0% to 15% in steps of 1%.

d. Plot a line graph of the table created in part c. above. [This graph is known as the *Operating Characteristic Curve* of the sampling plan.]

7. A company receives an order for five custom-made microchips at a price of $2500 each. The company will produce the chips one by one using a complex process which has only 67% chance of producing a defect-free chip at each trial. After five defect-free chips are produced the process will be stopped. The cost of production includes a $4800 fixed cost and a $870 variable unit cost. Thus if X number of chips are produced, the total cost of production would be $4800 + 870X$ dollars. The revenue minus the cost of production will be profit.

a. Find the expected value and variance of the profit.

b. Comment on the risk in the deal.

4. Normal Distribution

4.1 Introduction

The normal distribution is the most common continuous distribution occurring in practice. If the value of a random variable is affected by many independent causes, and the effect of each cause is small, then the random variable will closely follow a normal distribution. The length of a pin made by an automatic machine, the time taken by an assembly worker to complete an assembly, the weight of a baseball, the tensile strength of a bolt, and the volume of cola in a can are all normally distributed random variables. Much of the later chapters are based on the normal distribution. When a random variable X follows normal distribution with mean μ and variance σ^2, we write $X \sim N(\mu, \sigma^2)$.

For a normal distribution, the probability density function is

$$f(x) = \frac{1}{\sqrt{2\pi}\sigma} e^{\frac{1}{2}\left(\frac{x-\mu}{\sigma}\right)^2}.$$

A plot of the normal distribution with mean 100 and standard deviation 2 is shown in Figure 4.1.1.

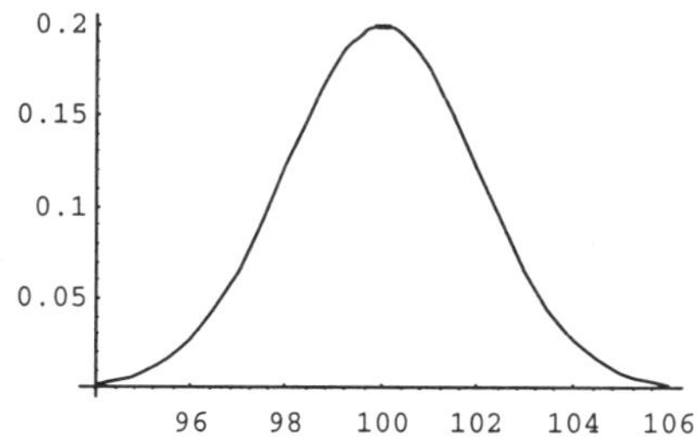

Figure 4.1.1. Normal Distribution

As seen in the figure, the normal distribution is symmetric with a (relative) kurtosis of 0, which means it is neither too peaked nor too flat.

We will be working with areas under the normal curve extensively. The spreadsheet formula used for the purpose is NORMDIST(which gives the area to the left of x under the normal curve with mean μ and standard deviation σ. The formula

`=NORMDIST(103,100,2,TRUE)`

gives the area to the left of 103 under the normal curve in Figure 4.1.1. The complement of that, namely,

`=1-NORMDIST(103,100,2,TRUE)`

will give the area to right of 103. The formula

`=NORMDIST(103,100,2,TRUE)-NORMDIST(99,100,2,TRUE)`

will give the area between 99 and 103. Note that the formula

`=NORMDIST(103,100,2,FALSE)`

will give the value of *f(x)* at x = 103, which we will almost never want.

Although the above formulas are powerful enough to carry out necessary computations, it is easier and even more powerful to work with the workbook named Normal Distribution. This template is shown in Figure 4.1.2, and as usual it can be used in conjunction with the **Goal Seek** command and the **Solver** tool to solve complicated problems.

As usual, the green cells are for input; the rest are protected. To use the template, make sure the right values are entered for Mean and Standard Deviation in cells B4 and C4. Cells B8 and C8 have been filled with x_1 and x_2 values so that cell D8 gives the probability $P(x_1 \le X \le x_2)$. Cell F8 contains 102 and therefore we get in cell G8 the area to the left of 102, and in cell H8 the area to the right of 102. Because cell J8 contains 0.9 we get in cell K8 that value of x to the left of which there will be 90% area. Because cell M8 contains 0.9 we get in cell N8 that value of x to the right of which there will be 90% area.

In later chapters, *confidence intervals* will be needed. A 2-tailed confidence interval $[x_1, x_2]$ is symmetric about the mean and contains an area equal to the confidence level denoted by $(1-\alpha)$. In the range F3:H5, we see 99%, 95% and 90% 2-tailed confidence intervals.

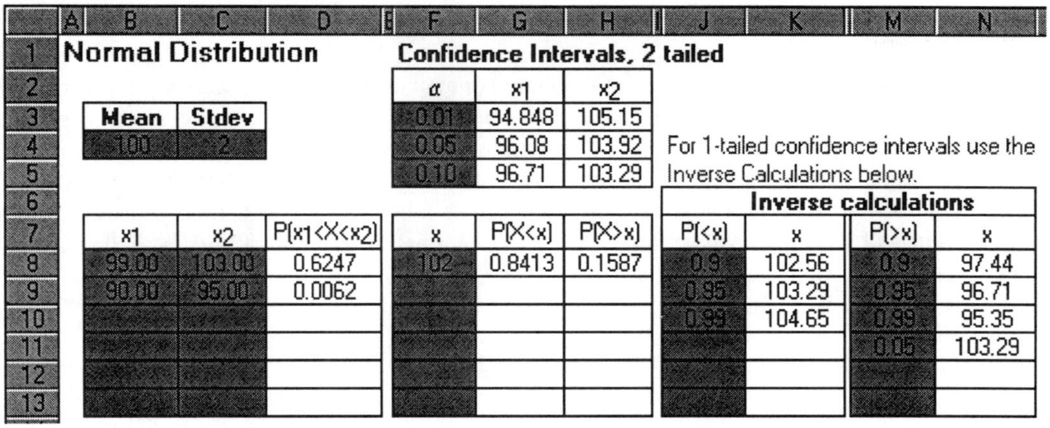

Figure 4.1.2 Normal Distribution Template
[Normal in Normal Distribution]

4.1.1 Standard Normal Distribution

A *standard normal distribution* is one for which the mean is zero and the standard deviation is 1. The standard normal random variable is denoted by Z. Thus $Z \sim N(0, 1^2)$. The formula

```
=NORMSDIST(z)
```

gives the area to the left of z in the standard normal distribution. Unlike NORMDIST, NORMSDIST needs only one argument. It clearly does not need the mean and standard deviation arguments since they are fixed at 0 and 1, respectively. The function NORMDIST takes the fourth TRUE or FALSE logical argument to return either the cumulative probability or the $f(x)$ value. The function NORMSDIST on the other hand, always returns the cumulative probability and does not take the logical argument.

Often, it is convenient to transform a normal random variable X into a standard normal random variable Z using the transformation

$$Z = (X - \mu)/\sigma$$

Note that this transformation gives the z-score of any x value as discussed in Chapter 1. The template in Figure 4.1.2 is applicable to the standard normal variable by setting the mean as zero and the standard deviation as 1.

4.2 Problem Solving

Most questions about normal random variables are very easy to answer using the template in Figure 4.1.2. For instance, suppose $X \sim N(100, 2^2)$. What is $P(99 \leq X \leq 103)$? Fill in cell B4 with 100 and cell C4 with 2. Fill in cell B8 with 99 and C8 with 103. The answer, 0.6247, appears in cell D8.

Some problems need the use of **Goal Seek ...** command. For instance, $X \sim N(\mu, 0.5^2)$; $P(X > 16.5) = 0.20$. What is μ? To answer this,

- Fill in cell C4 with 0.5 (because $\sigma = 0.5$)
- Fill cell B4 with some number, say 10, as your guess for μ
- Fill in cell F8 with `16.5` (because $P(X > 16.5)$ needs to be 0.20)
- Under the **Tools** menu, select the **Goal Seek ...** command.
- In the dialog box that appears, enter H8 in the **Set Cell** box.
- Enter 0.20 in the **To Value** box.
- Enter B4 in the **By Changing Cell** box.

The computer makes several trials and finds the answer as 16.08.

More complicated problems need the use of the **Solver** tool. For instance, consider the problem: $X \sim N(\mu, \sigma^2)$; $P(X > 28) = 0.80$; $P(X > 32) = 0.40$. What are μ and σ ?

One way to solve this problem is,

- Fill in cell B4 with 30 (which is a guessed value for μ)

- Fill in cell C4 with 2 (which is a guessed value for σ)
- Fill in cell F8 with 28
- Fill in cell F9 with 32
- Under the **Tools** menu select the **Solver**
- In the **Set Cell** box enter `H8`
- In the **To Value** box enter `0.80`
- In the **By Changing Cells** box enter `B4:C4`
- Click on the **Constraints** box and the **Add** button
- In the dialog box on the left hand side enter `H9`
- Select the = sign in the middle drop down box.
- Enter `0.40` in the right hand side box
- Click on **OK** button
- In the **Solver** dialog box that reappears, click on the Solve button.
- In the dialog box that appears at the end, select the **Keep Solver Solution** option

The Solver finds the correct values for the cells B4 and C4 that make cell G8 equal 0.80 and G9 equal 0.40. The answer shows up as $\mu = 31.08$ and $\sigma = 3.67$.

4.3 Normal Approximation of Binomial Distribution

When the number of trials n in binomial distribution is large (>1000) the calculation of binomial probabilities become difficult. Fortunately, the binomial distribution approaches the normal distribution as n increases and therefore we can approximate it as a normal distribution. This has been done in the sheet named "Normal approx. of Binomial" in the workbook named Normal Distribution Note that the mean is np and the standard deviation is $\sqrt{np(1-p)}$. The sheet is shown in Figure 4.3.1. The n and p parameters of the binomial distribution are input in cells B4 and C4 and they have been named `n` and `p`. The mean and the standard deviation of the corresponding normal distribution are calculated in cells F4 and G4. The formula in cell F4 is `=n*p`; and the formula in cell G4 is `=SQRT(n*p*(1-p))`. The rest of the sheet are identical to the normal distribution template.

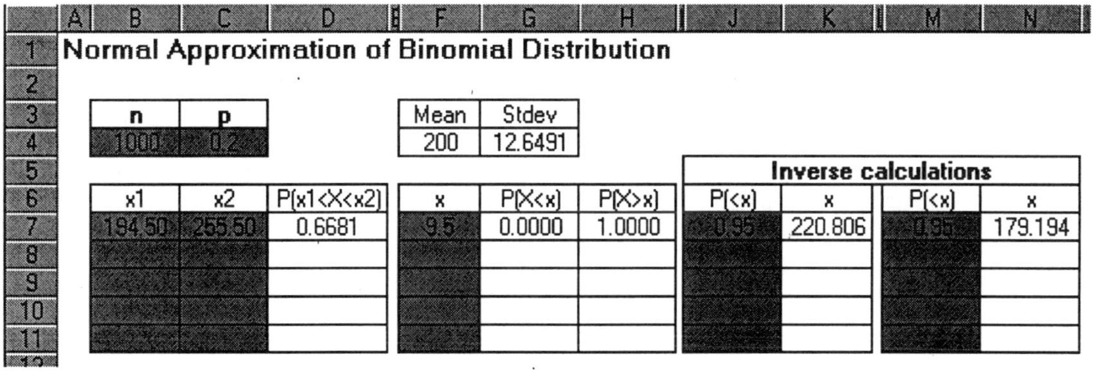

Figure 4.3.1 Normal Approximation of Binomial Distribution

Whenever a binomial distribution is approximated as a normal distribution, a *continuity correction* is required because binomial is discrete and normal is continuous. Thus, a column in the histogram of a binomial distribution for, say, $X = 10$ covers, in the continuous sense, the interval [9.5, 10.5]. Therefore, when we calculate the binomial probability of an interval, say, $P(195 \le X \le 255)$ we should subtract 0.5 on the left and add 0.5 on the right to get the corresponding normal probability, namely, $P(194.5 < X < 255.5)$. Adding and subtracting 0.5 in this manner is known as the correction for continuity. In Figure 4.3.1, this correction has been applied as seen in cells B7 and C7. Cell D7 has the binomial probability of $P(195 \le X \le 255)$.

4.4 Exercises

1. The GMAT scores of students who are potential applicants to a university are normally distributed with a mean of 487 and a standard deviation of 98.
i. What percentage of students will have scores exceeding 500?
ii. What percentage of students will have scores between 600 and 700?
iii. If the university wants only the top 75% the students to be eligible to apply, what should be the minimum GMAT score specified for eligibility?
iv. Find the narrowest interval that will contain 75% of the students' scores.
v. Find x such that the interval $[x, 2x]$ will contain 75% of the students' scores.

2. The profit (or loss) from an investment is normally distributed with a mean of $11,200 and a standard deviation of $8,250.
i. What is the probability that there will be a loss rather than profit?
ii. What is the probability that the profit will be between $10,000 and $20,000?
iii. Find x such that the probability that the profit will exceed x is 25%.
iv. If the loss exceeds $10,000 the company will be ruined. What is the probability of ruin?
v. Comment on the risk in the investment.

4.5 Projects

1. An automatic lathe produces pins whose lengths are normally distributed with a mean of 1.012" and a standard deviation of 0.018". The customer will buy only those pins with lengths in the interval [0.98, 1.02].

i. What percentage of the pins will be acceptable to the consumer?

ii. If the lathe can be adjusted to have the mean of the lengths to any desired value, what should it be adjusted to?

iii. Suppose the mean cannot be adjusted, but the standard deviation can be reduced, and it costs more and more to reduce the standard deviation further and further. What maximum value of the standard deviation would make 90% of the parts acceptable to the consumer? (Assume the mean to be 1.012.)

iv. Repeat question iii. with 95% and 99% of the pins acceptable.

v. In practice, which one do you think is easier to adjust, the mean or the standard deviation? Why?

vi. Assume it costs $\$150x^2$ to decrease the standard deviation by $(x/1000)$". Find the cost of reducing the standard deviation to the values found in questions iii. and iv.

vii. Now assume that the mean has been adjusted to the best value found in question ii. at a cost of $80. Calculate the reduction in standard deviation necessary to have 90%, 95% and 99% of the parts acceptable. Calculate the respective costs, per question vi.

viii. Based on your answers to questions vi. and vii. what is your recommended mean and standard deviation?

5. Sampling and Sampling Distributions

5.1 Introduction

When populations are large, it is difficult and too costly to measure every item in the population to calculate population parameters such as mean and variance. Instead, we take a random sample and infer population parameters from sample statistics. The word *parameter* is used to denote a characteristic of a population and *statistic* is used to denote a characteristic of a sample. A *random sample* is one where every item in the population is given equal chance of being included in the sample. All the results we are going to see are valid only if the sample is random.

The first thing to note is that a sample statistic is a random variable. For instance, if we take two different samples from the same population we would get two different sample means. Although the sample mean is random, intuition tells us that it will not be too different from the population mean. In this chapter, we shall see exactly how sample statistics will be distributed.

We denote the population size by N, population mean by μ, and population standard deviation by σ. We denote the sample size by n, sampled values by x_1, x_2, ..., x_n, sample mean by $\overline{X}$, and sample standard deviation by S. See Figure 5.1.1. With these notations we can say that $\overline{X}$ is an estimator of μ, and S is an estimator of σ.

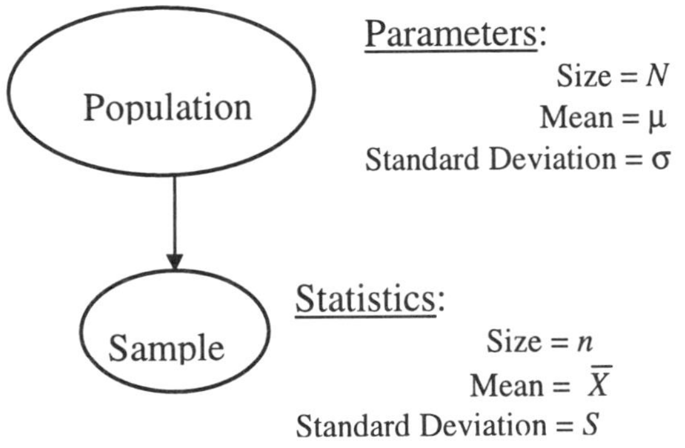

Figure 5.1.1 Notations

5.2 Sampling Distributions

By *sampling distribution,* we mean the distribution of a sample statistic when random samples are drawn from a population. We shall see the sampling distribution of three statistics, namely, the sample mean, the sample proportion and the sample variance,

5.2.1 Sample Mean

Suppose we take a random sample of size n from a population of size N. Consider the sum of the n items in the sample (call it sample sum). Because the sample is random, each item x in the population has n/N chance of being selected, and thus will contribute $(n/N)x$ to the sample sum. This makes the expected value of the sample sum equal (n/N) times the population sum which works out to $n\mu$. Dividing this by n we get the expected value of sample mean as μ. Because the expected value of sample mean is equal to the population mean, we call $\overline{X}$ an unbiased estimator of μ.

Next we shall consider the variance of $\overline{X}$. The variance of each sampled item is σ^2, and since they are all independent the variance of their sum is $n\sigma^2$. To get the variance of sample mean we divide this by n^2 and get σ^2/n. Thus the variance of sample mean is $1/n$-th of the population variance which shows that the variance will decrease as n increases.

The most striking result in sampling theory is that as n increases, the distribution of $\overline{X}$ will approach the normal distribution. This is known as the *Central Limit Theorem* (CLT), and it is the reason for predominant use of normal distribution in sampling theory. We have, by CLT, for large n, $\overline{X} \sim N(\mu, \sigma^2/n)$. For this purpose, an n value of 30 or more is considered large.

5.2.2 Sample Proportion

Suppose a large population consists of p proportion of successes and $(1 - p)$ proportion of failures. We take a random sample of size n and find x number of them to be successes. We thus calculate the proportion of successes in the sample (call it the sample proportion) denoted by $\hat{p}$ as x/n. What are the expected value and variance of $\hat{p}$? We note that x will be binomially distributed and therefore its mean is np and variance is $np(1 - p)$. Since $\hat{p} = x/n$, its expected value is p and variance is $p(1 - p)/n$. Thus $\hat{p}$ is an unbiased estimator of p.

5.2.3 Sample Variance

Coming to the sample variance S^2, the statistics $(n - 1)S^2/\sigma^2$ will follow a *chi-square distribution* with $(n - 1)$ degrees of freedom. Chi-square distribution is considered in detail in the next chapter.

5.3 Desirable Properties of Estimators

When the expected value of an estimator is equal to the true value, we call it an *unbiased* estimator. We have seen that both the sample mean and sample proportion are unbiased estimators.

An estimator is *efficient* if its variance is small. Among all unbiased estimators of μ, $\overline{X}$ is the most efficient. It is therefore the *best* estimator of μ.

An estimator is *consistent* if its probability of being close to the parameter it estimates increases as the sample size increases. Because $\overline{X}$ is unbiased and its variance, σ^2/n decreases as n increases, it is consistent.

An estimator is *sufficient* if it makes use of all the information in the data. Since $\overline{X}$ is calculated using all the sampled values, it is sufficient.

5.4 Exercises

To solve these exercises, use the Normal Distribution workbook.

1. Do problem 5-28 in Aczel's book.
2. Do problem 5-29 in Aczel's book.
3. Do problem 5-43 in Aczel's book.
4. Do problem 5-44 in Aczel's book.

6. Confidence Intervals

6.1 Introduction

Suppose the high temperature forecast for tomorrow is 50 degrees Fahrenheit. This does not mean it will be exactly 50 degrees; it could be slightly more or slightly less. A forecast is usually a statement, such as,

"There is a 95% chance that the high temperature tomorrow will be 50 ± 3 degrees."

In this case, we say that the 95% *confidence interval* for tomorrow's high temperature is 50 ± 3. We can also write this interval as [47, 53]. The central value of 50 is the *point estimate* and 3 is the *half-width* of the confidence interval.

Compare the above forecast with the next,

"There is a 96% chance that the high temperature tomorrow will be 50 ± 3.2 degrees."

This forecast is better in that the confidence has increased, but worse in that the interval has widened. Consider the next,

"There is a 99.99% chance that the high temperature tomorrow will be 50 ± 30 degrees."

Although there is almost 100% confidence, this forecast is useless because the wide interval of ± 30 degrees does not help us get a reasonable idea about what the weather is going to be. Thus, we would like a confidence interval to be as narrow as possible and have as much confidence as possible. In the calculation of confidence intervals, the objective is to minimize the half-width and maximize the confidence level at the same time. This is usually done by fixing the confidence level at any one of the three standard values 90%, 95% or 99%, and then calculating the narrowest interval that has that confidence level.

In this chapter, we shall study how to calculate confidence intervals for population means and proportions based on sample statistics.

6.2 Confidence Intervals for Population Means

6.2.1 The basic case

The symbol used for confidence level is $(1 - \alpha)$, so that a 95% confidence level will mean an α of 5%. In a standard normal distribution (z-distribution), if we want the narrowest interval that captures an area of $(1 - \alpha)$, it should be symmetric about the center because the distribution peaks at the center. We should leave an area of $\alpha/2$ on each tail so that the central interval would cover an area of $(1 - \alpha)$. The z value that has an area of $\alpha/2$ to the right is denoted by the symbol $z_{\alpha/2}$. See Figure 6.2.1. The $(1 - \alpha)$ confidence interval is thus $[-z_{\alpha/2}, z_{\alpha/2}]$.

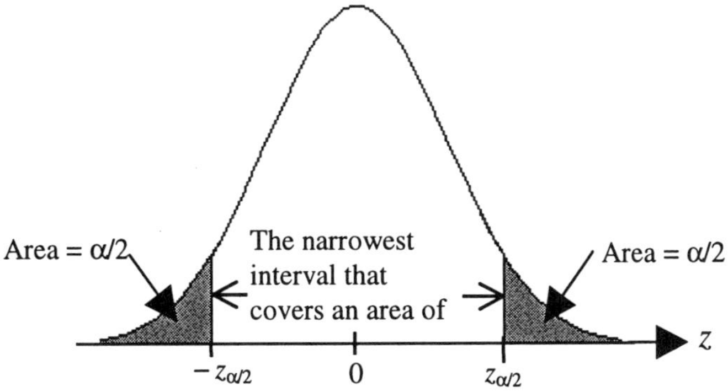

Figure 6.2.1 A $(1 - \alpha)$ Confidence Interval

6.2.2 One-sided Confidence Intervals

At times, we want to be confident only about the maximum or the minimum possible value. We then do not care about the width of the interval. In this case, we need a *one-sided confidence interval*. We leave the whole area α on one of the two tails and the rest of the z-axis becomes the $(1 - \alpha)$ confidence interval, which would be either $[-\infty, z_\alpha]$ or $[z_\alpha, \infty]$.

6.2.3 Finite Population Correction

In most cases, the populations is large or infinite and the variance of the sample mean, $V(\bar{x})$ equals σ^2/n. This value is used in confidence interval calculations. When the population is finite, $V(\bar{x})$ is a tiny bit less than σ^2/n. This difference grows as N keeps decreasing and becomes comparable to n. Indeed, if $N = n$, then the whole population is sampled and the sample mean must equal the population mean, implying a variance of zero. In general, if n/N is more than 1%, a finite population correction is necessary. $V(\bar{x})$ reduces by a factor of $(N - n)/(N - 1)$ and the width of the confidence interval decreases by a factor of $\sqrt{(N-n)/(N-1)}$. This correction also applies to confidence intervals for population proportions.

6.2.4 When Population Standard Deviation is Known

The calculation of confidence intervals for population means from sample means depends upon whether or not the population standard deviation σ is known. Most of the time, σ is not known; but if it is known it would help us to decrease the half-width for a given confidence level. We shall first see the case where σ is known.

From the Central Limit Theorem, we know that if either the population is normally distributed, or the sample size n is more than 30, then the sample mean $\bar{x} \sim N(\mu, \sigma^2/_n)$. Therefore a $(1 - \alpha)$ confidence interval for $\bar{x}$ is $\mu \pm z_{\alpha/2}*\sigma/\sqrt{n}$. This means that with $(1 - \alpha)$ confidence, we can say the *difference* between μ and $\bar{x}$ is not more than $z_{\alpha/2}*\sigma/\sqrt{n}$. Therefore, a $(1 - \alpha)$ confidence interval for μ is

$$\bar{x} \pm z_{\alpha/2}*\sigma/\sqrt{n}.$$

In the Confidence Intervals workbook, the page named "CI for mu" can be used to calculate confidence intervals for μ. This page is shown in Figure 6.2.1. In cell D8, the function CONFIDENCE has been used.

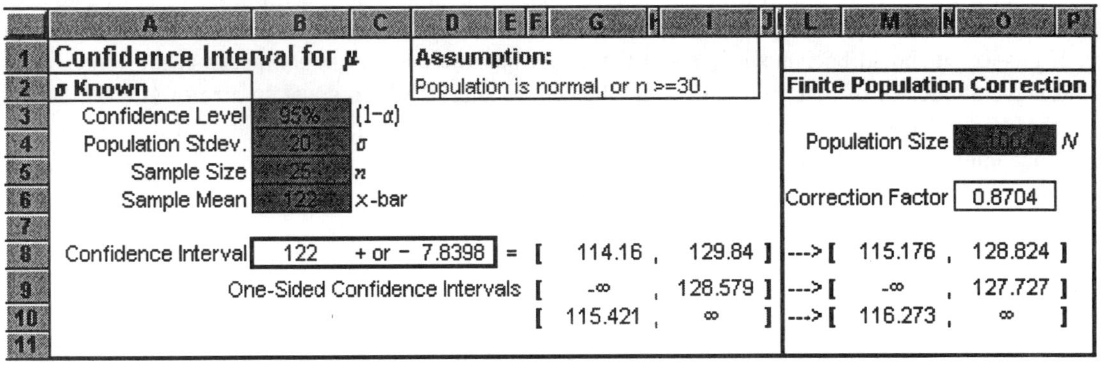

Figure 6.2.1. Confidence Interval for μ
[CI for mu in Confidence Intervals]

The formula `=CONFIDENCE(α,σ,n)` returns the half-width $z_{\alpha/2}*\sigma/\sqrt{n}$. The formula in cell D8 is

`=CONFIDENCE((1-B3),B4,B5).`

When the data for α, σ and n are entered in cells B3, B4 and B5, the required confidence interval is displayed in the range B8:D8.

The same template contains one-sided confidence intervals as well. The formula in cell I9 is

`=B6+CONFIDENCE(2*(1-B3),B4,B5).`

where the α value has been doubled so that an area of α will go on one tail. The formula in cell I10 is

`=B6-CONFIDENCE(2*(1-B3),B4,B5).`

To apply finite population correction to confidence intervals, the population size N must be entered in cell O4. The correction factor is computed in cell O6 with the formula =SQRT((O4-B5)/(O4-1)). The corrected confidence intervals appear in the range L8:P10.

6.2.5 When Population Standard Deviation is Unknown

Most of the times, the population standard deviation σ is not known, and the population may be known to have a normal distribution. We then substitute the sample standard deviation s in its place and use the *t-distribution* to calculate confidence intervals. The t-distribution was discovered by W. S. Gossett, who wrote under the pen name Student. It is therefore called Student's t-distribution. Gossett proved that when a random sample is drawn from a *normally distributed population*, the statistic $\left[\dfrac{\bar{x} - \mu}{s / \sqrt{n}}\right]$ follows the t-distribution with $(n - 1)$ degrees of freedom, which helps us calculate confidence intervals for μ after knowing $\bar{x}$.

The t-distribution is similar to the standard normal distribution (the z-distribution) in that it has a mean of zero and is symmetric; but it is flatter with larger variance (> 1) and smaller (relative) kurtosis (< 0). Unlike the z-distribution, of which there is only one, there is one t-distribution for each number of degrees of freedom, df = 1, 2, ... As the df increases, the variance approaches 1, the (relative) kurtosis approaches zero, and the t-distribution approaches the z-distribution.

In the Confidence Intervals workbook, there is a page for t-distribution. It is very similar to the one for normal distribution, with an input cell for df. All calculations similar to those of normal distribution are possible on this template. Two Excel functions that relate to the t-distribution are TDIST and TINV. Their format is somewhat non-standard and needs to be noted carefully. The formula

```
=TDIST(ABS(t),df,1)
```

will give the tail area to the right of ABS(t). The reason for taking the absolute value of t using the ABS function is that TDIST does not accept a negative value. The third argument of 1 denotes the number of tails. If it is 2 then twice the area, accounting for both tails, would be returned.

The formula

```
=TINV(α,df)
```

will return $t_{\alpha/2}$ which is that value of t that has $\alpha/2$ area to its right. Note that the TINV function *always* assumes a 2-tailed case. If you need t_α, then the formula to use is

```
=TINV(2*α,df).
```

The half-width of the confidence interval for μ when σ is not known is given by $t_{\alpha/2} * s/\sqrt{n}$, where the df for the t value is $(n - 1)$. On the bottom half of the page named "CI for mu" in Confidence Intervals workbook the confidence interval for μ when σ is unknown is calculated. See Figure 6.2.2. When the input data is entered in the range B14:B17, the confidence interval is displayed in the range B19:D19. The formula used in cell D19 is

```
=TINV(1-B14,B15-1)*B17/SQRT(B15).
```

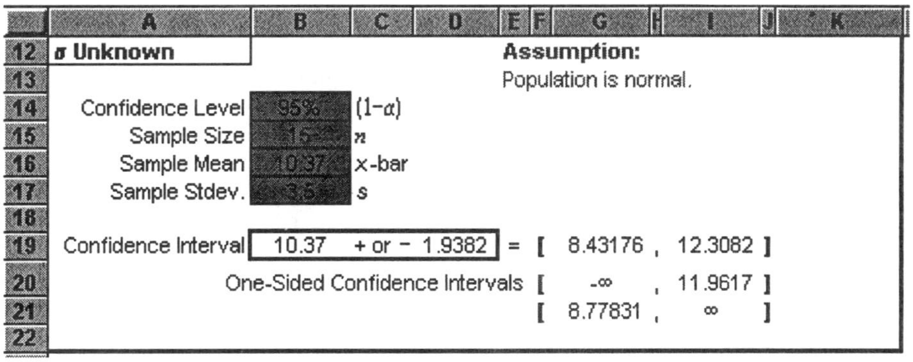

Figure 6.2.2 CI for μ When σ is Unknown
[CI for mu in Confidence Intervals]

The one-sided confidence intervals are found in the range F20:J21. The formula in cell I20 is

```
=B16+TINV((1-B14)*2,B15-1)*B17/SQRT(B15).
```

And the formula in cell G21 is

```
=B16-TINV((1-B14)*2,B15-1)*B17/SQRT(B15).
```

6.3 Confidence Intervals for Population Proportions

When np and $n(1-p)$ are both at least 5, the $(1-\alpha)$ confidence interval for population proportion p is given by

$$\hat{p} \pm z_{\alpha/2}\sqrt{\hat{p}(1-\hat{p})/n}$$

where $\hat{p}$ is the sample proportion. This formula has been implemented in the template shown in Figure 6.3.1.

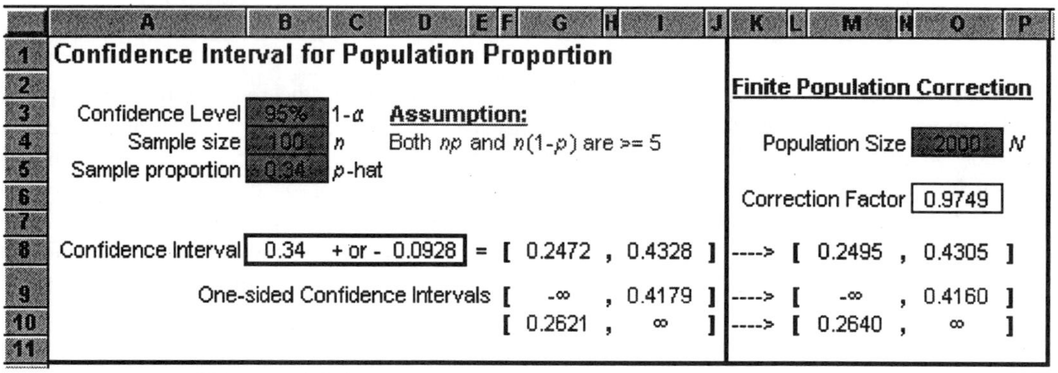

Figure 6.3.1. Confidence Interval for Population Proportion
[CI for p in Confidence Intervals]

When the input data are entered in the range B3:B5, the confidence interval appears in the range B8:D8 and in the range F8:J8. One-sided confidence intervals appear in the range F9:J9. If finite population correction is needed, the population size N should be entered in cell O4, and the confidence intervals appear in the range L8:P10.

6.4 Confidence Intervals for Population Variance

A well known result in sampling theory is that if random samples are drawn from a <u>normal</u> population, the statistic $(n-1)*S^2/\sigma^2$ where S^2 is the sample variance and σ^2 is the population variance would follow a *chi-square distribution* with $(n-1)$ degrees of freedom. Letting z denote a standard normal random variable, chi-square with k degrees of freedom is the sum of k independent z^2 values. A chi-square distribution has its range from zero to infinity, mean equal to the degrees of freedom, variance equal to twice the degrees of freedom and is positively skewed. The sheet named "Chi-Square Distn." in Confidence Intervals is similar to the t-distribution page and can be used similarly. The Excel functions CHIDIST and CHIINV are useful while working with the chi-square distribution. The formula `=CHIDIST(`χ^2`,`*df*`)` will return the area to the right of the χ^2 value entered as the first argument. The formula `=CHIINV(`α`,`*df*`)` will return that χ^2 value that has α area to its right. In both cases, the second argument designates the degrees of freedom.

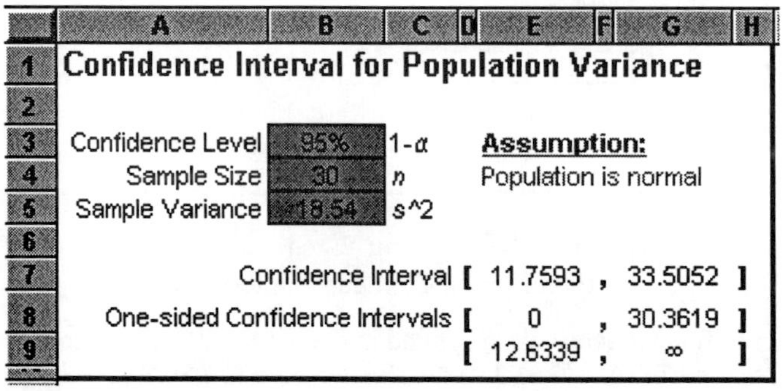

Figure 6.4.1 Confidence Interval for Population Variance
[CI for Popn. Variance in Confidence Intervals]

When samples are drawn from a <u>normal</u> population, the $(1 - \alpha)$ confidence interval for σ^2 is given by

$$\left[\frac{(n-1)s^2}{\chi^2_{\alpha/2}}, \frac{(n-1)s^2}{\chi^2_{1-\alpha/2}} \right]$$

where the chi-square has $(n - 1)$ degrees of freedom. Figure 6.4.1 shows the sheet named "CI for Popn. Variance" in Confidence Intervals, which implements the above formula using the CHIINV function. After the input data are entered in the range B3:B5, the confidence intervals appear in the range D7:H9.

6.5 Sample Size Determination

An important practical decision in confidence interval estimation is how large the sample should be. Too small a sample will yield too wide an interval; too large a sample will require too much effort. The objective in sample size determination is to find the smallest sample that will yield the desired width for the confidence interval.

In the case of population mean, we have seen that the half-width for the confidence interval (denote it by B) is given by $B = z_{\alpha/2}\, \sigma/\sqrt{n}$. Rearranging this to bring n to the left hand side yields the formula for minimum n as,

$$\text{Minimum } n = \left\lceil \left[\frac{z_{\alpha/2}\sigma}{B} \right]^2 \right\rceil$$

where the symbol $\lceil\ \rceil$ means rounding up. For instance, $\lceil 48.2 \rceil = 49$. Rounding up is necessary since n cannot be a fraction. The excel formula ROUNDUP implements this operation; in other words, the formula `=ROUNDUP(48.2,0)`, where the second argument 0 specifies the number of decimals, will return 49. Note that the formula for minimum n calls for σ. Since the population mean itself is not known, it is unlikely that σ will be known. One should make a guess, or use past experience with similar data to enter a reasonably good value for σ. Figure 6.5.1 shows the sheet named "Sample Size" in Confidence Intervals. When the input data are entered in the range B5:B7, the minimum sample size appears in cell B9. The formula in cell B9 is,

```
=ROUNDUP((NORMSINV((1-B5)/2)*B7/B6)^2,0).
```

The currently input values are from Aczel's Example 6-6.

	A	B	D	E
1	**Sample Size Determination**			
2				
3	**For Population Mean, μ**		**For Population Proportion, p**	
4				
5	Confidence Level	95%	Confidence Level	99%
6	Half-Width	120	Half-Width	0.1
7	Population Stdev. (guessed)	400	Population Proportion (guessed)	0.25
8				
9	Minimum Sample Size	43	Minimum Sample Size	125

Figure 6.5.1 Minimum Sample Size Calculation
[Sample Size in Confidence Intervals]

In the case of population proportion, the half width $B = z_{\alpha/2} \sqrt{p(1-p)/n}$. Rearranging this to get n on the left hand side yields,

$$\text{Minimum } n = \left\lceil \left(\frac{z_{\alpha/2}}{B} \right)^2 p(1-p) \right\rceil .$$

In the template shown in Figure 6.5.1, when the input data are entered in the range E5:E7 the minimum sample size appears in cell E9. The formula in E9 is,

```
=ROUNDUP((NORMSINV((1-E5)/2)/E6)^2*E7*(1-E7),0).
```

The currently input values are from Aczel's Example 6-7.

6.6 Exercises

1. The lengths of pins produced by an automatic lathe are normally distributed. A random sample of 20 pins gives a sample mean of 0.992" and a sample standard deviation of 0.013".

i. Give a 95% confidence interval for average lengths of all pins produced.

ii. If it is claimed that the lathe has been set to have $\mu = 1 \pm 0.002$" could you reject it with 95% confidence?

iii. Give a 99% confidence interval for the average lengths of all pins produced.

2. You take a random sample of 100 pins from the lot supplied by a vendor, and test them. You find 3 of them defective.

i. What is the 95% confidence interval for % defective in the whole lot?

ii. If the vendor claims that the lot contains not more than 5% defectives, can the claim be rejected with 95% confidence?

iii. What is the maximum % defective in the whole lot, with 99% confidence?

6.7 Projects

1. It is desired to estimate the average length of pins produced by an automatic lathe to within 0.002" with 95% confidence level. σ is guessed to be 0.015".

i. What is the minimum sample size?

ii. If the value of σ may be anywhere between 0.010" and 0.018", tabulate the minimum sample size required for σ values from 0.010" to 0.018" in steps of 0.002".

iii. If the cost of sampling and testing *n* pins is (25 + 6**n*) dollars, tabulate the costs for the same σ range as in question ii. above.

2. It is desired to estimate the % defective in a lot of pins supplied by a vendor to within 1% with 90% confidence level. The actual % defective is guessed to be 4%.

i. What is the minimum sample size?

ii. If the actual % defective may be anywhere between 3% and 6%, tabulate the minimum sample size required for actual % defective from 3% to 6% in steps of 0.5%.

iii. If the cost of sampling and testing *n* pins is (25 + 6*n*) dollars, tabulate the costs for the same % defective range as in question ii. above.

3. A company wants to conduct a telephone poll to estimate the proportion of voters who favor a particular candidate in a presidential election, to within 2% error with 95% confidence. It is guessed that the proportion is 53%.

i. What is the minimum sample size?

ii. The actual proportion may be anywhere from 40% to 60%. Construct a 2-dimensional table for the minimum sample size required with half-width ranging from 1% to 3% in steps of 1% along the rows, and actual proportion ranging from 40% to 60% in steps of 5% along the columns.

iii. Inspect the table produced in question ii. above. Comment on the relative sensitivity of the minimum sample size to the actual proportion and to the desired half-width. Also find the worst case value for the actual proportion.

iv. If the cost of polling n people is (250 + 0.6*n*), tabulate the cost as in question ii. above.

7. Hypothesis Testing

7.1 Introduction

Suppose a vendor has shipped the pins you had ordered, and claims that the lot contains not more than 4% defectives, per agreement. The agreement is that if the lot contains not more than 4% defectives, you would buy the lot and make the payment. Typically, the number of parts in a lot will be too numerous to conduct a 100% inspection to determine the % defective in the lot. Sampling is the solution, and because the sampling is meant for accepting or rejecting a lot, the process is called *Acceptance Sampling*. To test if the lot contains not more than 4% defectives, suppose you sample one pin at random and test it.

Case 1: The sampled pin is defective. You reject the lot because the sampled pin was defective. Is everything all right? No. The lot may indeed contain less than 4% defective and therefore be acceptable. You just happened to sample a defective pin, and are probably committing the error of rejecting an acceptable lot. This type of error is known as *Type I error.*

Assume that the lot contains 2% defectives. Then the probability of the sampled pin being defective is 2% and thus there is a 2% chance that you are committing a Type I error. If the plan is to sample one pin at random and reject the lot if the sampled pin is defective, we note: When the lot contains 2% defectives, there would be 2% chance of Type I error; if it contains 3% defectives, there would be 3% chance of Type I error... In practice, though, we do not know the % defective in the lot and therefore do not know the exact probability of Type I error. But we can say that *at worst*, the probability of Type I error is 4% which occurs when the actual % defective is 4% and the lot is therefore just acceptable. This value of 4%, which is the (maximum) probability of Type I error is known as the *p-value* of the evidence that the sampled pin is defective. The *p*-value is also known as the *significance level.* For making a rejection decision based on a sample evidence, it is customary to place a ceiling on the allowable *p*-value of that evidence. This ceiling on *p*-value is denoted by α, and it is usually 5%. Other popular values for α are 1% and 10%. The complement of α, or $(1 - \alpha)$, is known as the *confidence level.* Because a Type I error immediately affects the producer of the pins, α is also known as the (maximum allowable) *producer's risk.*

Does the plan of sampling one pin at random and rejecting the lot if the sampled pin is defective satisfy $\alpha = 5\%$? We just saw that the *p*-value of the evidence (of the sampled pin being defective) is 4%, which is less than α. Therefore the plan is satisfactory.

Case 2: The sampled pin is not defective. You accept the lot because the sampled pin was good. Is everything all right? No. The lot may indeed contain more than 4% defective and therefore be unacceptable. You just happened to sample a good pin, and are probably committing the error of accepting a bad lot. This type of error is known as *Type II error.*

Assume that the lot contains 7% defectives. Then the probability of the sampled pin being good is as high as 93% and thus there is 93% chance that you are committing a Type II error. If the plan is to sample one pin at random and accept the lot if the sampled pin is good, we note: When the lot contains 7% defectives, there would be a 93% chance of committing a Type II error; if the lot contains 6% defectives, there would be a 94% chance of Type II error... Since we do not know the actual % defective in the lot, we do not know the exact probability of Type II error. At worst, it is 96% which occurs when the % defective is minutely larger than 4% and the lot is just unacceptable. This 96% is the complement of the *p*-value we saw in Case 1. It is customary to specify a ceiling on the probability of Type II error for a given % defective in the lot, such as: "When there are 8% defectives in the lot, the probability of Type II error should not exceed 10%." This ceiling of 10% is denoted by β. The complement of β, or $(1 - \beta)$, is known as the *power* of the sampling plan. Because a Type II error immediately affects the consumer of the pins, β is also knows as the *consumer's risk.*

Does the plan of sampling one pin at random and accepting the lot if the sampled pin is good satisfy $\beta = 10\%$ when % defective = 8%? Let us see. If there are 8% defectives, then the probability of the sampled pin being good and you accepting the lot is as high as 92%. Consequently, this sampling plan is far from satisfactory. [The solution is to base the acceptance decision on a much larger sample.]

The discussion above explains several important concepts of hypothesis testing, namely, Type I and Type II errors, *p*-value or significance level of an evidence, α, confidence level, β and power. We will be using them repeatedly.

An acceptance/rejection decision is a delicate balance between Type I and Type II errors. For a given sample size, if we adjust the plan so as to reduce the chances of one type of error, we would inevitably increase the chances of the other. Increasing the sample size would decrease the chances of both types of error, but, of course, it would increase the time, effort and money needed to execute the plan. The overall objective in hypothesis testing is to limit the chances of Type I and Type II errors to desired levels using the minimum sample size.

The desired levels for the chances of Type I and Type II errors depend on how serious each type of error is. Often, one type of error is more serious than the other and hence would need to be controlled more closely. If the pins are to be used to mount engines on a commercial airplane, accepting a bad lot would be a grave Type II error. On the other hand, if they are used to mount pictures on walls, Type I error may be costlier.

Because acceptance/rejection decisions can have serious consequences, we formalize them by writing down the exact claim that we are testing. The claim is stated as the *Null Hypothesis* and is denoted by H_0. The counterclaim, which would be the exact opposite, is called the *Alternative Hypothesis* and is denoted by H_1 or H_a. We shall use H_1. For the example above,

H_0: % defective is $\leq 4\%$

H_1: % defective is $> 4\%$

In general, if the null hypothesis is true, no special action would be necessary; if the alternative hypothesis is true, some special action would be necessary.

7.2 Tests of Population Mean

Consider an automatic machine which bottles cola into 2-liter (2000 cc) bottles. Because of changes in working conditions, wear and tear, and variations in the process itself, the exact amount put into bottles will vary. So the machine needs to be checked periodically to ensure that it puts 2000cc on the average into each bottle. Three different cases are possible here based on three different concerns, namely, consumer protection, profit and process control. We shall see them one by one.

Consumer protection requires the average amount to be at least 2000 cc so that consumers get their money's worth. Thus the hypotheses would be formulated as

$$H_0: \mu \geq 2000$$
$$H_1: \mu < 2000$$

To test the null hypothesis, let us say a random sample of 49 bottles is taken, and the bottles are tested for the exact amounts of their contents. From sampling theory, we know that since the sample is large the sample mean, $\overline{X}$, will be normally distributed with mean μ and variance σ^2/n. If the computed sample mean, $\bar{x}$, is greater than 2000 there is nothing to complain about and the null hypothesis (H_0) would not be rejected. But if $\bar{x}$ is less than 2000 there is reason to doubt H_0, and the lesser it gets the more doubtful we become. At some point we might consider $\bar{x}$ to be too low to accept H_0, and reject it. Because the rejection of H_0 occurs when $\bar{x}$ is low, which corresponds to the left tail of the normal distribution of $\overline{X}$, we call this case a *1-tailed test with rejection on the left.*

In rejecting the null hypothesis, we might be committing a Type I error. The probability of this occurring, though, is not as clear cut as in the case of Acceptance Sampling (with $n = 1$) that we saw in the previous section. What is the p-value here? The convention is to declare the area to the left of $\bar{x}$ on the normal distribution for $\overline{X}$ as the p-value. *The p-value of an evidence should therefore be interpreted as the probability of the evidence being as unfavorable to* H_0 *as, or more unfavorable to* H_0 *than, it currently is.* It is also customary to calculate the p-value on the standard normal distribution after converting $\bar{x}$ into its z-score using the formula

$$z = \frac{\bar{x} - \mu}{\sigma / \sqrt{n}}.$$

Specifically, the p-value will be the area under the standard normal curve to the left of the computed z value. Since the statistic z decides the outcome of the test, it is called the *test statistic*. Once again, the maximum allowable p-value is α, and its complement, $(1 - \alpha)$, is known as the confidence level. Whenever the p-value is less than α, H_0 is rejected.

Let us carry out a sample calculation. We shall assume that from past experience σ is known to be 7. Suppose $\bar{x}$ is 1998, then the test statistic $z = (1998 - 2000)/(7/\sqrt{49}) = -2$. The p-value is then calculated on the spreadsheet using the formula `=NORMSDIST(-2)` which yields 0.0228. Thus H_0 will be rejected if α is 5% or 10%, but not if it is 1%.

Figure 7.2.1 shows the sheet named "Tests of mu" in Hypothesis Testing workbook. When the input data are entered into the shaded cells of this template, the test statistic z and the p-values appear in the designated places.

	A	B	C	D	E	F	G
1	**Hypothesis Testing - Population Mean**						
3				**σ Known**			
4	Hypothesized μ	2000	μo		**Assumption**		
5	Sample size	49	n	Either	1. Population Normal		
6	Sample Mean	1998	x-bar	Or	2. Sample Size > 30		
7	Population Stdev.	7	σ				
8					**2-tailed**	**1-tailed Left**	**1-tailed Right**
9	z	-2.0000		p-value	0.0455	0.0228	0.9772
12				**σ Unknown**			
13	Hypothesized μ	65	μo		**Assumption**		
14	Sample size	21	n		Population Normal		
15	Sample Mean	62.5	x-bar				
16	Sample Stdev.	3	s				
17					**2-tailed**	**1-tailed Left**	**1-tailed Right**
18	t	-3.8188		p-value	0.0011	0.0005	0.9995

Figure 7.2.1 Hypothesis Test for μ
[Tests of mu in Hypothesis Testing]

The formula for calculating the test statistic z in cell B9 is

`=(B6-B4)/(B7/SQRT(B5)).`

The data entered currently in the template corresponds to the example calculation above. The p-value for the 1-tailed test with rejection on the left, 0.0228, appears in cell F9.

When the population is finite, a finite population correction can be applied. The population size N should be entered in cell J5. The corrected z and p-values appear in their places (not shown in Figure 7.2.1). Note that the p-value has decreased. Thus a hypothesis that is not rejected when the correction was not applied may be rejected when the correction is applied.

Often, the population standard deviation σ is unknown. In this case, we can conduct a t-test if the population is normal. In a t-test, we substitute the sample standard deviation s in place of σ. The statistic

$$t = \frac{\overline{X} - \mu}{S/\sqrt{n}}$$

follows Student's t-distribution with $(n - 1)$ degrees of freedom. A t-distribution with degrees of freedom = df has mean zero, variance $df/(df - 2)$ for $df > 2$, and is symmetric. As the df increases it approaches the normal distribution.

The bottom half of the template seen in Figure 7.2.1 contains the t-test. Note that the input area contains s instead of σ. The formula in cell B18 for calculating t is similar to the one in cell B9. This test is also a 1-tailed test with rejection on the left. The p-value of 0.0005 is less than the specified $\alpha = 1\%$; hence the null hypothesis is rejected.

The second concern with the automatic bottling machine is the profit motive. Putting more than 2000 cc on the average into each bottle would waste cola and reduce profits. A special action, of stopping and resetting the machine, is necessary when the average is greater than 2000. Hence the hypotheses will be formulated as

$$H_0: \mu \leq 2000$$
$$H_1: \mu > 2000$$

Here the rejection will occur when the sample mean is too much more than 2000 or when the test statistic z or t goes too far into the right tail. This case is therefore a *1-tailed test with rejection on the right*. On the template shown in Figure 7.2.1, the *p*-value for this case appears in cell G9 or G18 must be read-off.

The third concern with the bottling machine is that of process control. To be in control, the average amount bottled should be neither too much nor too little. It should be as near 2000 as possible. The special action of stopping and resetting the machine is necessary on both tails of the test statistic. Here the hypotheses will be formulated as

$$H_0: \mu = 2000$$
$$H_1: \mu \neq 2000$$

Since the rejection of H_0 occurs on both tails, this test is called a *2-tailed test*. On the template, the *p*-value for this case appears in cell E9 or E18.

7.3 Tests of Population Proportion

7.3.1 *z*-Test

A *z*-test can be used to test hypotheses about population proportion *p*. The test statistic *z* is calculated using the formula

$$z = \frac{\hat{p} - p_0}{\sqrt{p_0(1-p_0)/n}}$$

where $\hat{p}$ is the sample proportion and p_0 is the hypothesized value for *p*. The test may be 1-tailed or 2-tailed. The sheet named "z-Test for p" in Hypothesis Testing workbook is shown in Figure 7.3.1.

	A	B	C	D	E	F	G
1	**z-Test for Population Proportion**						
4	Hypothesized *p*	0.7	*p*		**Assumption**		
5	Sample size	210	*n*		Both np and n(1-p) >= 5		
6	Sample Proportion	0.6190	*p*-hat				
7					**2-tailed**	**1-tailed Left**	**1-tailed Right**
8	z	-2.5599		*p*-value	0.0105	0.0052	0.9948
11	**Correction for Finite Population**						
12							
13	Population Size	1000	*N*				
14					**2-tailed**	**1-tailed Left**	**1-tailed Right**
15	z	-2.8787		*p*-value	0.0040	0.0020	0.9980

Figure 7.3.1 *z*-Test for Population Proportion
[*z*-Test for p in Hypothesis Testing]

When the input data are entered in the shaded cells, *z* and *p*-values appear in their designated places. The bottom portion is for applying finite population correction.

If the sample proportion is known straight away it is directly entered in cell B6. At times, we know the number of successes *x* in the sample and not the sample proportion. In this case we enter a formula in cell B6 to calculate the population proportion. In the example problem $x = 130$. Hence the formula =130/210 has been entered in cell B6.

For a 2-tailed test the *p*-value is 0.0105. At an α of 5% the null hypothesis is rejected.

7.3.2 Binomial Test

The use of *z*-test is an approximation under the condition that both *np* and $n(1 - p)$ are at least 5. On the spreadsheet, we really do not need to approximate and the condition is unnecessary. We can assume

x to be binomially distributed and conduct a more accurate binomial test. The sheet named "Binomial Test for p" in Hypothesis Testing workbook is shown in Figure 7.3.2.

	A	B	C	E	F	G	H
1	**Binomial Test for *p***						
3					**2-Tailed**	**1-Tailed Left**	**1-Tailed Right**
4	Hypothesized *p*	0.72	*p*	*p*-value	0.0147	0.0073	0.9927
5	Sample size	210	*n*				
6	#Successes	130	x				
7	Sample proportion	0.6190	*p*-hat				

Figure 7.3.2 Binomial Test for Population Proportion
[Binomial Test for p in Hypothesis Testing]

On this template, when the input data is entered in the shaded cells, the p-values appear in their places. The data currently entered is for a 2-tailed test. The p-value is therefore 0.0147, and the null hypothesis is rejected at an α of 5%. The p-value calculated here is more accurate than that seen in Figure 7.3.1.

7.4 Tests of Population Variance

	A	B	C	D	E	F	G
1	**χ^2 Test for σ^2**						
2					**2-Tailed**	**1-Tailed Left**	**1-Tailed Right**
3	Hypothesized Popn. Variance	1	σ^2	*p*-value	0.03451	0.98274	0.01726
4	Sample Size	31	*n*		**Assumption:**		
5	Sample Variance	1.62	s^2		Population Normal		
7	χ^2	48.6					

Figure 7.4.1 Testing σ^2
[Test for Popn. Var. in Hypothesis Testing]

When random samples are drawn from a *normally distributed* population, the statistic $(n-1)s^2/\sigma^2$ follows a χ^2 distribution with $(n-1)$ degrees of freedom. Thus a χ^2 test can be carried out for testing hypotheses regarding σ^2. Figure 7.4.1 shows the sheet named "Test for Popn. Var." in Hypothesis Testing workbook. When the input data are entered into the shaded cells, the χ^2 value and the p-values appear in their designated places. The currently entered data is for a 1-tailed test on the right. The p-value is therefore 0.01726 and the null hypothesis is rejected.

7.5 Calculating the Power of a Test

7.5.1 Why Power is important

Most hypothesis tests are conducted with primary attention given to Type I error, which is controlled through the specification of α. However, Type II errors are also costly, at times even costlier than Type I error. It is therefore necessary to keep a watch on the chances of Type II error.

The probability of an error (Type I or II) depends upon the true value of the population parameter. Suppose we are testing μ, and the null hypothesis is $\mu \leq 100$. The probability of Type II error then will depend upon the actual μ. A typical statement about Type II error would then say, "When $\mu = 102$, the probability of Type II error is 0.1." If μ increases further, the null hypothesis becomes "more false" and thus becomes more likely to be rejected; hence the probability of Type II error will decrease. The power of a test is the complement of the probability of Type II error.

7.5.2 Power of a μ Test

Figure 7.5.1 shows the sheet named "Power of a mu Test" in Hypothesis Testing workbook. On this template, the input data are entered in the shaded cells. In addition, the type of test is selected in the drop down box. The template assumes that the critical value(s) for the sample mean is (are) decided according to the α value in cell B6. For particular values of actual μ, the probability of Type II error and the Power are calculated in the range E4:I5. As seen in the range E3:E5, when μ = 7.7, the power of the test is 0.9218.

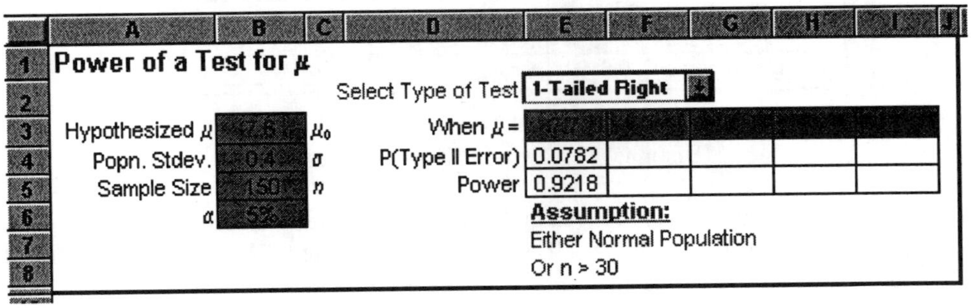

Figure 7.5.1 Power of a μ Test
[Power of a mu test in Hypothesis Testing]

7.5.3 Power of a *p* Test

Figure 7.5.2 shows the sheet named "Power of a p test" in Hypothesis Testing workbook. Its use is similar to that of the previous template. When the input data are entered in the shaded cells and the type of test is selected from the drop down box, the results appear in the range E4:I5.

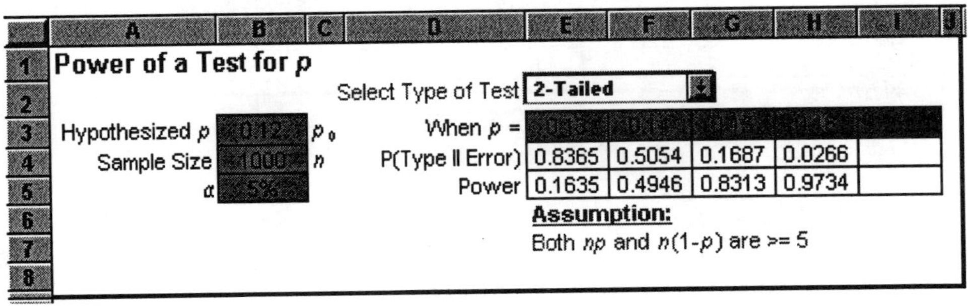

Figure 7.5.2 Power of a *p* Test
[Power of a p test in Hypothesis Testing]

7.6 Sample Size Determination

7.6.1 Introduction

An important practical decision in hypothesis testing is sample size determination. The objective of hypothesis testing is to limit the chances of Type I and Type II errors to specified maximums and at the same time minimize the sample size. Sample size determination is thus an optimization problem with Type I and Type II error constraints. These constraints will be specified as, "under such and such condition the chance of Type I/II error should not exceed such and such %."

The Solver tool is ideal for solving problems of this type. The fundamentals of using the Solver tool are explained in Chapter 0.

7.6.2 Optimal Sample Size for Testing μ

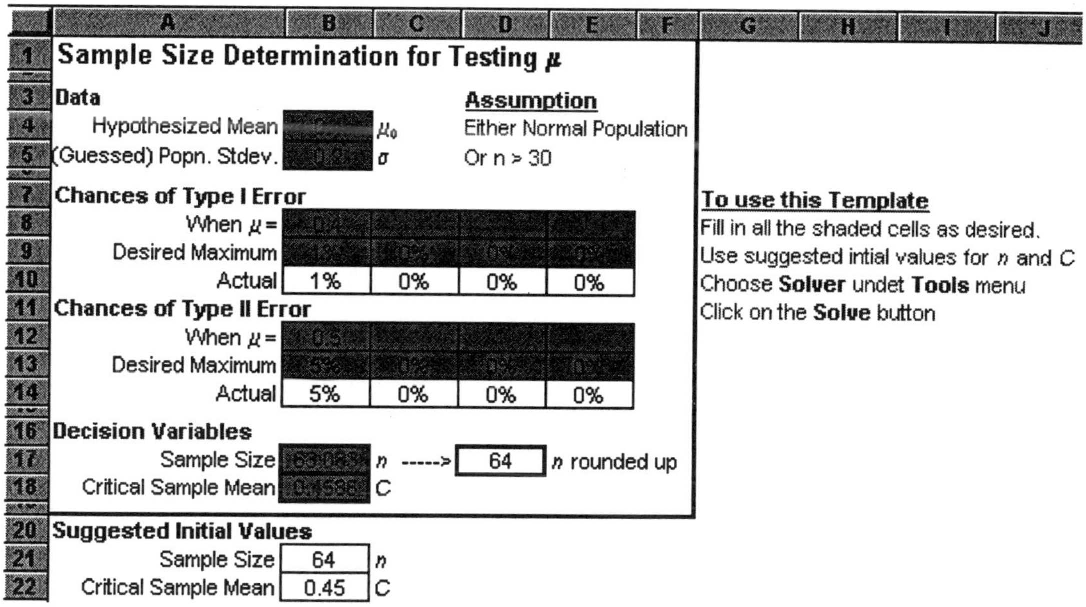

Figure 7.6.1 Determining n to Test μ
[n fo rTesting mu in Hypothesis Testing]

The sheet named "n for Testing mu" in Hypothesis Testing workbook is shown in Figure 7.6.1. This template uses the Solver for optimizing the sample size. The Solver settings are shown in Figure 7.6.2.

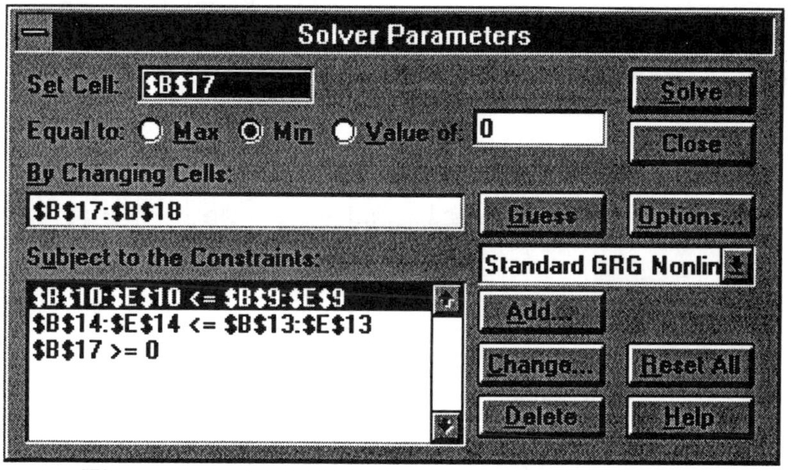

Figure 7.6.2 Solver Settings for Determining n to Test μ

In this setting, **Set Cell** B17 **Equal to Min** specifies that we want to minimize the sample size n in cell B17. **By changing cells** B17 and B18 tells the Solver to adjust the sample size n and the critical value of $\bar{x}$, denoted by C, beyond which the null hypothesis will be rejected. Three constraints have been listed in the constraints area. The first states that the actual probabilities of Type I Error should be less than the desired levels. The second is similar, and corresponds to Type II Error. The third ensures that the sample size does not stray into negative values. If the constraints regarding Type I and Type II Errors are meaningful and reasonable, pressing the **Solve** button will find an optimal solution. If the constraints are inconsistent, Solver would not be able to find a solution and would display a message to that effect.

Some points to note while using this template are:

1. If there is only one constraint each for Type I and Type II Errors, then the suggested value for n in the cell B21 itself would be optimal. The Solver can then be used to find the optimal C. The formula for optimal n under these circumstances is

$$\text{Minimum } n = \left\lceil \left(\frac{(|z_0| + |z_1|)\sigma}{\mu_0 - \mu_1} \right)^2 \right\rceil$$

where z_0 is the critical z implied by the Type I Error constraint when $\mu = \mu_0$, and z_1 and μ_1 similarly correspond to Type II Error constraint. The symbol $\lceil\ \rceil$ means rounding up to the nearest integer. The spreadsheet formula in cell B21 of the template, which implements this calculation of optimal n is

```
=ROUNDUP(((NORMSINV(1-B9)+NORMSINV(1-B13))*B5/(B8-B12))^2,0).
```

Because there is only once constraint each for Type I and Type II errors, the suggested value of 64 is optimal for n. If there are more constraints, the Solver would have to be used.
2. The objective in the current setup is to minimize n. A problem may have a total cost function based on Type I and Type II Error costs which may have to be minimized.
3. If the Solver is unable to find an optimal solution, check all the data input, especially the Type I and Type II Error constraints. Make sure n and C have been set to the suggested values. After any correction, re-run the Solver.

7.6.3 Optimal Sample Size for Testing *p*

The case of finding the optimal sample size for testing p is very similar to the case of m seen in the previous section. Figure 7.6.3 shows the sheet named "n for Testing p" in Hypothesis Testing workbook. After entering the necessary data in the shaded cells, one may use the Solver to find the solution.

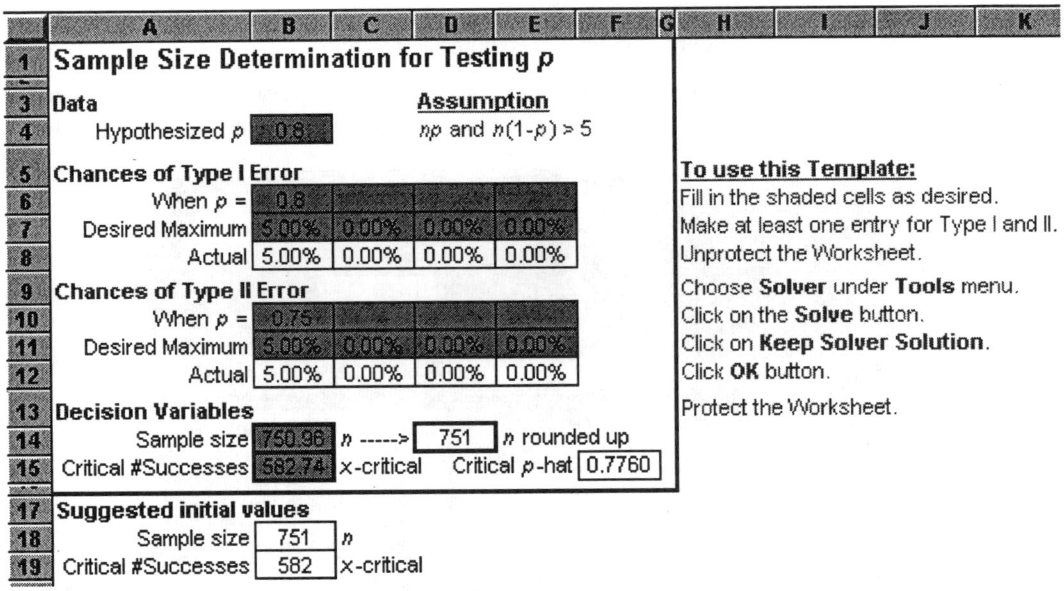

Figure 7.6.3 Optimal *n* for Testing *p*
[N for Testing p in Hypothesis Testing]

While using this template, the following points may be noted.
1. If there is only one constraint each for Type I and Type II Errors, then the suggested value for n in the cell B18 itself would be optimal. The Solver can then be used to find the optimal value for x-critical. The formula for optimal n under these circumstances is

$$\text{Minimum } n = \left\lceil \left(\frac{|z_0|\sqrt{p_0(1-p_0)} + |z_1|\sqrt{p_1(1-p_1)}}{p_0 - p_1} \right)^2 \right\rceil$$

where z_0 is the critical z implied by the Type I Error constraint when $p = p_0$, and z_1 and p_1 similarly correspond to Type II Error constraint. The symbol $\lceil\ \rceil$ means rounding up to the nearest integer. The spreadsheet formula in cell B18 of the template, which implements this calculation of optimal n is

```
=ROUNDUP(((NORMSINV(1-B7)*SQRT(B6*(1-B6))+NORMSINV(1-B11)*SQRT(B10*(1-
B10)))/(B6-B10))^2,0)
```

Because there is only once constraint each for Type I and Type II Errors, the suggested value of 751 is optimal for n. If there are more constraints, the Solver would have to be used.

2. The objective in the current setup is to minimize n. A problem may have a total cost function based on Type I and Type II Error costs which may have to be minimized.

3. If the Solver is unable to find an optimal solution, check all the data input, especially the Type I and Type II Error constraints. Make sure n and x-critical have been set to the suggested values. After any correction, re-run the Solver.

4. A necessary assumption here is that np and $n(1 - p)$ are both at least 5. At times this may not be satisfied. Specially in Acceptance Sampling problems, the proportion p which stands for the proportion of defectives will be very small and np may very well be less than 5. The template is then not applicable, and one has to use the sheet named "n for Testing p Binomial" in the same workbook. The disadvantage in using this binomial template is that the Solver will not work because binomial distribution is discrete. One has to solve the problem manually by adjusting n and x-critical. While solving a problem manually like this, one may note that changing the value of x-critical will change Type I and Type II Error probabilities greatly, and changing n will change them only slightly.

7.7 Exercises

1. Solve Exercise 7-1 in Aczel's book.
2. Solve Exercise 7-9 in Aczel's book.
3. Solve Exercise 7-16 in Aczel's book.
4. Solve Exercise 7-17 in Aczel's book.
5. Solve Exercise 7-21 in Aczel's book.
6. Solve Exercise 7-24 in Aczel's book.
7. Solve Exercise 7-37 in Aczel's book.
8. Solve Exercise 7-44 in Aczel's book.
9. Solve Exercise 7-45 in Aczel's book.
10. Solve Exercise 7-50 in Aczel's book.
11. Solve Exercise 7-53 in Aczel's book.
12. Solve Exercise 7-56 in Aczel's book.
13. Solve Exercise 7-82 in Aczel's book.

7.8 Projects

1. A producer and a consumer of pins want to find a sampling plan that would be acceptable to both of them. The producer wants a lot that has only 1% defectives to have at least 99.5% probability of acceptance, and a lot that has 2% defectives to have at least 98% probability of acceptance. The consumer wants a lot that has 8% defectives to have at most 10% probability of acceptance, a lot that has 10% defectives to have at most 2% probability of acceptance and a lot that has 12% defectives to have not more than 0.5% chance of acceptance. Find the optimal value for sample size n and acceptance number c (x-critical).

2. A sampling plan has sample size $n = 100$ and acceptance number $c = 3$. The probability distribution of the % defective (p) in incoming lots is given by the table below.

p	Prob
1%	0.3
2%	0.4
3%	0.2
4%	0.1

i. When a random lot is received, what is its probability of acceptance under the given sampling plan? [*Hint: Construct a joint probability table.*]

ii. Given that a lot is accepted, what is the probability that it has 1% defectives? 2% defectives?

iii. What is the expected % defective in a lot accepted under this sampling plan? [This quantity is known as the Average Outgoing Quality (AOQ).]

8. The Comparison of Two Populations

8.1 Introduction

While comparing two population parameters, we need to make use of the fundamental result that if X and Y are two *independent* random variables, then

$$E[X - Y] = E[X] - E[Y]$$
$$V(X - Y) = V(X) + V(Y).$$

These two equations help us in finding the sampling distribution for differences in population means and proportions.

8.2 The Analysis ToolPak macros

Excel contains Analysis ToolPak macros for conducting the five different tests described in this chapter, namely, the *z*-test, three types of *t*-tests and the *F*-test. You may use them to conduct these tests, but there are some disadvantages in using them. The disadvantages are:

1. When the data is updated, the results will not update automatically. You have to run the macro again. Also, the **Goal Seek** command and the **Solver** tool cannot be used.
2. The Analysis ToolPak macros will work only with raw data. If you know the mean and variance of the samples but not the actual raw data, you cannot use these macros.
3. Besides the test results, you may want to see other details such as confidence intervals. This is possible only in the templates.

The use of an Analysis ToolPak macro is described in Chapter 1, Section 1.4. This chapter does not contain any description of the five macros mentioned above.

8.3 Comparing Two Population Means

In order to compare two population means, it is necessary to take two *random and independent* samples from the two populations. Independence is achieved by making the selection of one sample not influence the selection of the other in any way. We shall use the notations shown in Figure 8.3.1.

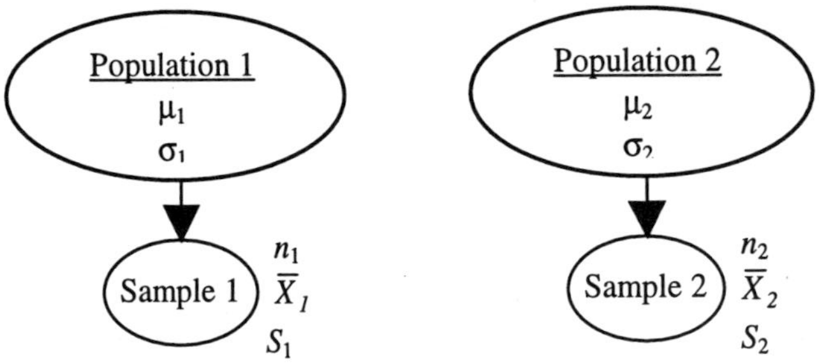

Figure 8.3.1 Comparing Two Population Means

The hypotheses are about $\mu_1 - \mu_2$, such as

$$H_0: \mu_1 - \mu_2 = D$$
$$H_1: \mu_1 - \mu_2 \neq D$$

for a 2-tailed test where D is the hypothesized d$_{if}$ference in the means. Often, D is zero. For a 1-tailed test with rejection on the left the hypotheses will be

$$H_0: \mu_1 - \mu_2 \geq D$$
$$H_1: \mu_1 - \mu_2 < D$$

and for rejection on the right tail

$$H_0: \mu_1 - \mu_2 \leq D$$
$$H_1: \mu_1 - \mu_2 > D$$

8.3.1 The z-Test: Population Variances Known

From sampling theory, we note that

$$E[\overline{X}_1 - \overline{X}_2] = \mu_1 - \mu_2$$

$$V(\overline{X}_1 - \overline{X}_2) = V(\overline{X}_1) + V(\overline{X}_2) = \sigma_1^2/n_1 + \sigma_2^2/n_2.$$

By the Central Limit Theorem, as n_1 and n_2 both increase, $(\overline{X}_1 - \overline{X}_2)$ will approach the normal distribution. When σ_1 and σ_2 are known, we can therefore conduct a z-test if either n_1 and n_2 are both more than 30 or the two populations are normal. The z statistic is given by the formula

$$z = \frac{(\overline{X}_1 - \overline{X}_2) - D}{\sqrt{\dfrac{\sigma_1^2}{n_1} + \dfrac{\sigma_2^2}{n_2}}}$$

If we want a $(1 - \alpha)$ confidence interval for the difference in the population means, it will be

$$(\bar{x}_1 - \bar{x}_2) \pm z_{\alpha/2}\sqrt{\frac{\sigma_1^2}{n_1} + \frac{\sigma_2^2}{n_2}}\,.$$

The sheet named "z-Test" in Testing Differences, shown in Figure 8.3.2 is the template to use here. After entering the input data in the shaded cells, the p-values of hypothesis tests and the confidence intervals appear at the designated places.

Note: The user may have to calculate the sample means and then input them into cells B6 and E6. If the raw data from the samples are entered in another area of the sheet and are named "Sample1" and "Sample2", then the formula `=AVERAGE(Sample1)` may be entered in cell B6, and the formula `=AVERAGE(Sample2)` in cell E6. One may also use the COUNT function to calculate the sample sizes. For instance, the formula `=COUNT(Sample1)` would return n_1.

While entering population variance, one should make sure that it is the variance (and not standard deviation). The data currently entered are from Example 8-3 of Aczel. The standard deviation of population 1 is given as 212. The actual entry in cell B8 is the formula `=212^2`, and similarly the formula in cell E8 is `=185^2`. The test being 2-tailed, the p-value is 0.0000 and therefore the null hypothesis is rejected.

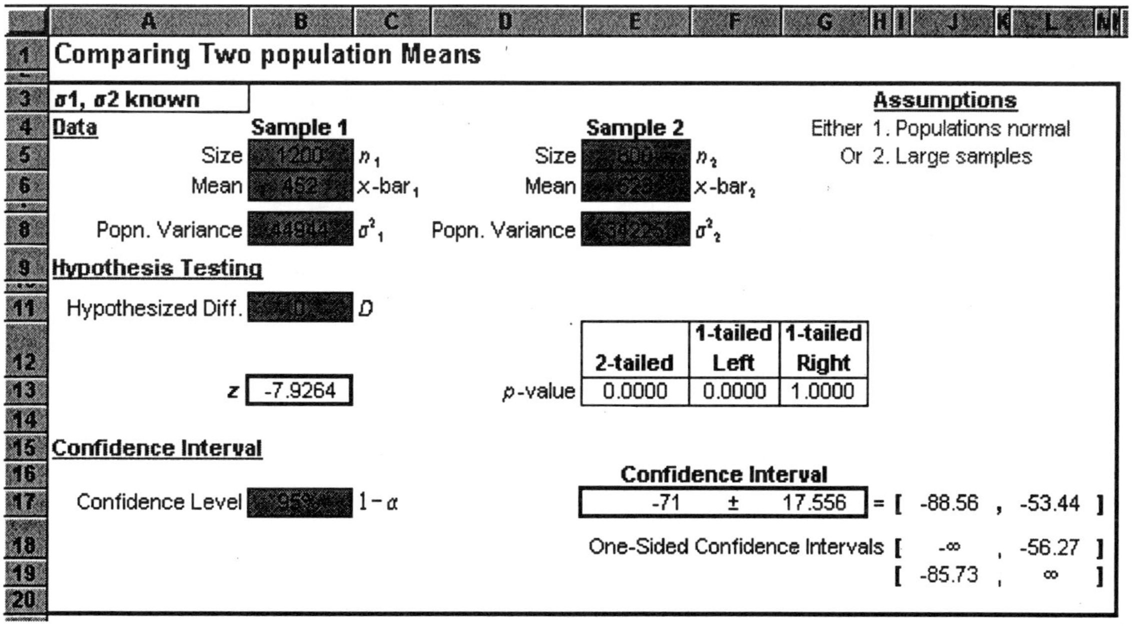

Figure 8.3.2. z-Test for Difference in Population Means
[z-Test in Testing Differences]

8.3.2 The *t*-Test: Population Variances Unknown

The *z*-test is possible only when the population variances are known, but very often they are unknown. If they are unknown, a *t*-test is possible. Here there are two cases, one where the population variances are assumed to be equal and the other where they are assumed to be unequal. The assumption of equal variances is justified when the causes for variations are the same for both populations. For instance, we may be testing the difference between the average strengths of pins produced from two different raw materials but using the same machinery and process. It is reasonable in this case to assume equal variances. One may also conduct an *F*-test described in Section 8.5 to confirm that they are equal.

8.3.2.1 When Population Variances Are Equal

When population variances are equal, a pooled variance is calculated by pooling the results of both samples, and used in the calculations of test statistic or confidence intervals. The formula for pooled variance s_p^2 is

$$s_p^2 = \frac{(n_1 - 1)s_1^2 + (n_2 - 1)s_2^2}{n_1 + n_2 - 2}$$

The sheet named "t-Test Equal Variance" is shown in Figure 8.3.3. When the input data are entered in the shaded cells, *p*-values and confidence intervals appear in the designated places. In this case, besides sample means, sample variances also must be calculated and entered. If raw data of the two samples are entered in another part of the sheet and named "Sample1" and "Sample2", then in cell B7 the formula `=VAR(Sample1)` may be entered, and in cell E7 the formula `=VAR(Sample2)` may be entered.

The *t* statistic is calculated using the formula

$$t = \frac{(\bar{x}_1 - \bar{x}_2) - D}{\sqrt{s_p^2\left(\frac{1}{n_1} + \frac{1}{n_2}\right)}}$$

The corresponding formula in cell B13 is

```
=((B6-E6)-B10)/SQRT(B12*(1/B5+1/E5)).
```

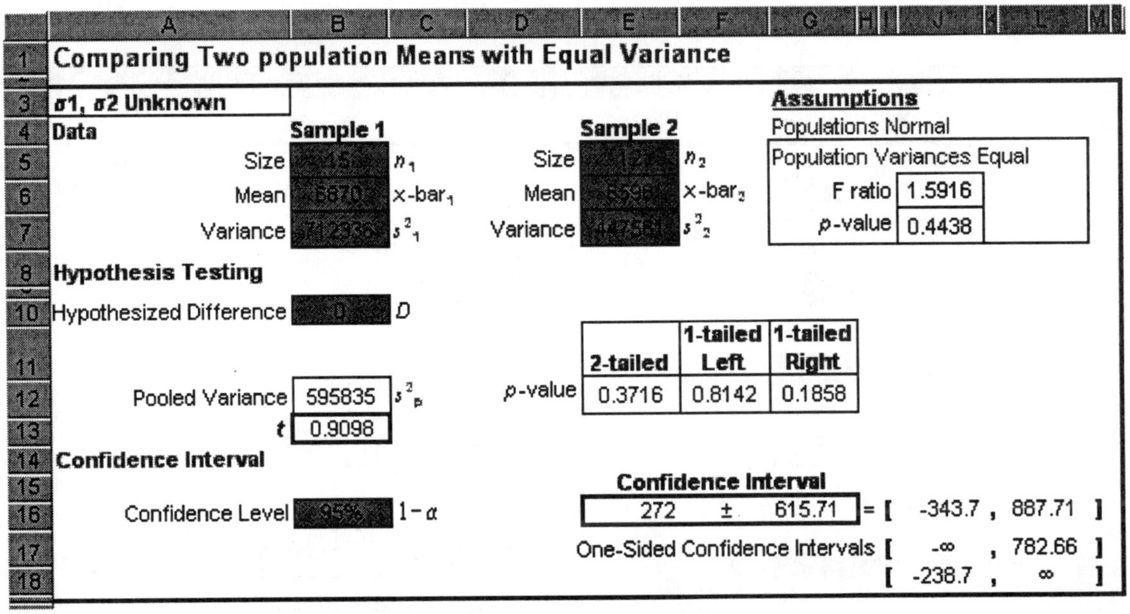

Figure 8.3.3. *t*-Test for Equal Population Variance
[t-Test Equal Variance in Testing Differences]

To test how good the assumption of equal variances is, the *F*-ratio is calculated in cell J6 and the corresponding *p*-value is calculated in cell J7. If the *p*-value is too small (say, less than 0.2), the assumption is questionable; the hypothesis may be repeated assuming unequal variances as described in the next section. For the currently entered data taken from Aczel's Example 8-6, the assumption seems acceptable.

Reading the problem (Example 8-6 of Aczel) shows that the test is a 1-tailed test with rejection on the right. The *p*-value therefore is 0.1858; the null hypothesis cannot be rejected even at an α of 10%.

8.3.2.2 When Population Variances Are Unequal

When the population variances are assumed unequal a pooled variance is not calculated. Instead the two sample variances are used separately in the formula for *t*.

$$t = \frac{(\bar{x}_1 - \bar{x}_2) - D}{\sqrt{\left(\frac{s_1^2}{n_1} + \frac{s_2^2}{n_2}\right)}}$$

The sheet named "t-Test Unequal Variance" in Testing Differences, shown in Figure 8.3.4, implements this scheme.

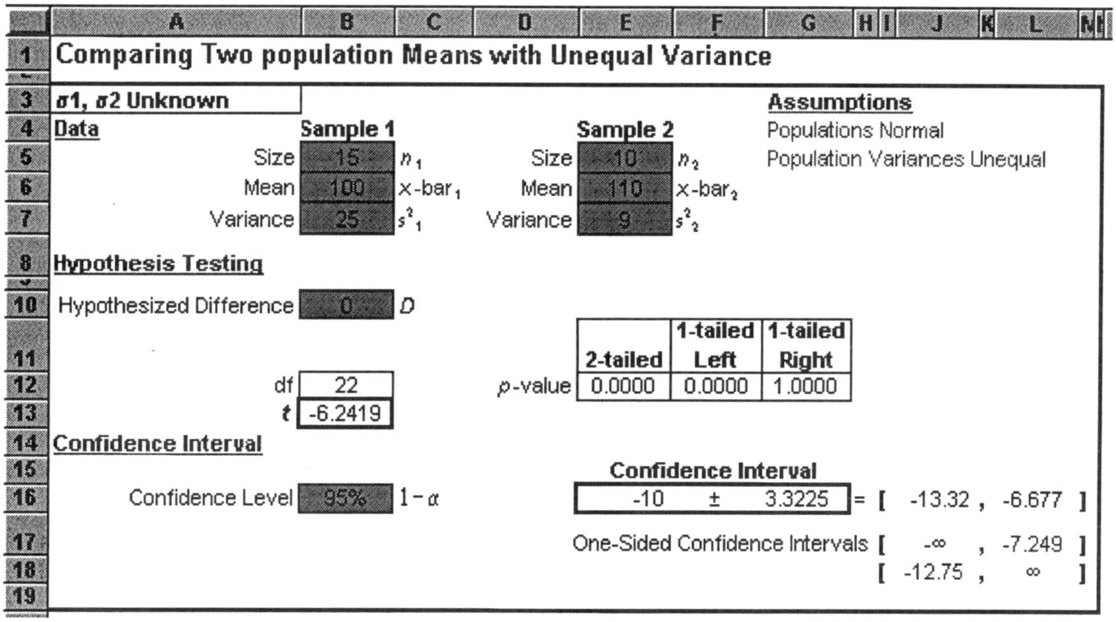

Figure 8.3.4. t-Test with Unequal Population Variances
[t-Test Unequal Variance in Testing Differences]

The formula to calculate the test statistic *t* in cell B13 is

```
=((B6-E6)-B10)/SQRT(B7/B5+E7/E5)
```

The degrees of freedom *df* for the *t* statistic is given by the formula

$$df = \left\lfloor \frac{(s_1^2/n_1 + s_2^2/n_2)^2}{\frac{(s_1^2/n_1)^2}{n_1 - 1} + \frac{(s_2^2/n_2)^2}{n_2 - 1}} \right\rfloor .$$

The brackets $\lfloor \; \rfloor$ signify that the quantity inside should be rounded down to the nearest integer. Rounding down the degrees of freedom gives the benefit of doubt to the null hypothesis. The corresponding spreadsheet formula in cell B12 is

```
=INT((B7/B5+E7/E5)^2/((B7/B5)^2/(B5-1)+(E7/E5)^2/(E5-1))).
```

The INT function which stands for "integral part" does the rounding down. The function ROUNDDOWN could have also been used in its place.

The data currently entered in the template is for a 2-tailed test. Since the p-value is almost zero, the evidence is extremely significant; the null hypothesis is rejected.

8.3.3 Paired Difference

At times it might be possible to pair the observations in the two samples and take the difference in each pair of observations. Usually this happens when subjects are exposed to a treatment, measurements are taken before and after the treatment, and these measurements are compared to test the effectiveness of the treatment. In effect, this amounts to a single population test where the population is the set of all possible differences in the measurements.

A paired difference test is more efficient than the other tests because it has less chances of Type I and Type II errors for the same sampling effort. The efficiency obtains because when measurements are made on the same subject before and after a treatment, effects of extraneous variables such as age, race and gender on the before/after difference are avoided. Thus the measured difference due to the treatment is more accurate and reliable. Hence, whenever a paired difference test is possible, one should settle for that rather than for other types of tests.

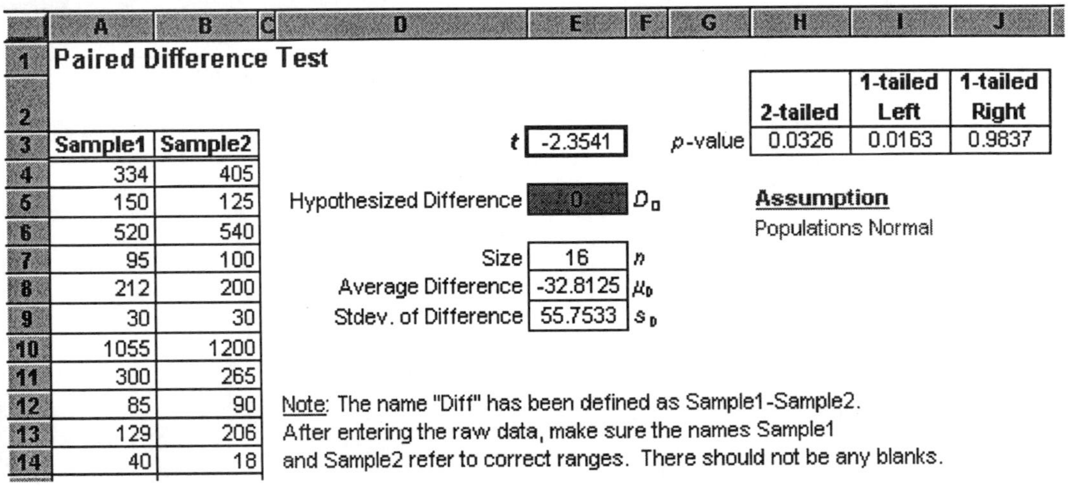

Figure 8.3.5. t-Test for Paired Difference
[t-Test Paired Difference in Testing Differences]

Figure 8.3.5 shows the template. The data currently input is for a 1-tailed test with rejection on the left. The p-value is 0.0163. At an α of 5% the null hypothesis is rejected; at an α of 1% it cannot be rejected.

8.4 Comparing Two Population Proportions

Comparing two population proportions can be done using a z-test if the two samples are sufficiently large. Here a large sample means both np and $n(1 - p)$ are at least 5.

From sampling theory we know that the mean of a sample proportion $\hat{p}$ is population proportion p and the variance is $p(1 - p)/n$. Thus the difference between two sample proportions will have a mean of $p_1 - p_2$ and a variance of $p_1 (1 - p_1)/n_1 + p_2 (1 - p_2)/n_2$. But we do not know p_1 and p_2 during a hypothesis test. If the two population proportions are hypothesized to be different, then we substitute $\hat{p}_1$ *and* $\hat{p}_2$ in their places. If the two population proportions are hypothesized to be equal, then we substitute a pooled proportion $\hat{p}$ in both places. This is analogous to the use of s_1^2 and s_2^2 separately or combining them into s_p^2 in the case of t-test for comparing population means. The pooled proportion is given by the formula

$$\hat{p} = \frac{x_1 + x_2}{n_1 + n_2}.$$

Figure 8.4.1 shows the sheet named "Popn. Proportion" in Testing Differences, which can be used for this test. After the input data are entered in the range B4:B5 and E4:E5, the test statistic z appears in B11 when the hypothesized difference is zero. The corresponding p-values appear in the range E10:G10. The value of z is given by the formula

$$z = \frac{\hat{p}_1 - \hat{p}_2}{\sqrt{\hat{p}(1-\hat{p})(\frac{1}{n_1}+\frac{1}{n_2})}}$$

where $\hat{p}$ is the average proportion.

The formula for $\hat{p}$ in cell B10 is `=(B5+E5)/(B4+E4)` and the formula for z in cell B11 is

`=(B6-E6)/SQRT(B10*(1-B10)*(1/B4+1/E4)).`

The confidence intervals appear in the range I19:M21.

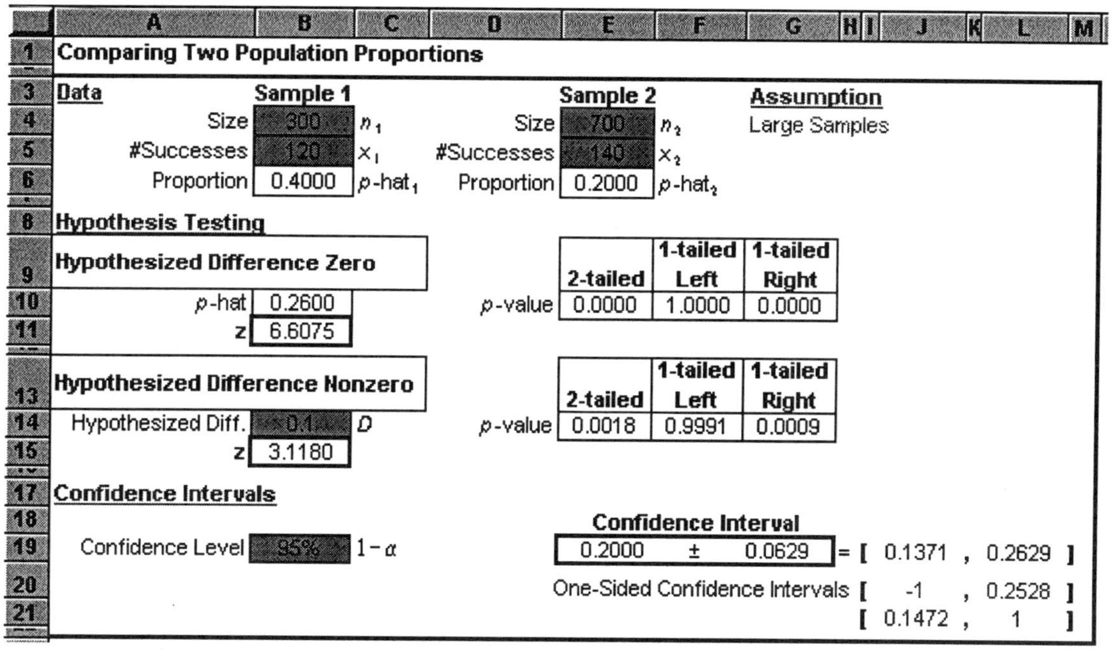

Figure 8.4.1 z-Test for Difference in Population Proportions
[Popn. Proportions in Testing Differences]

If the hypothesized difference is not zero, then it is entered in cell B14. The corresponding z and p-values appear in cell B15 and in the range E14:G14 respectively. The data entered currently is for a 1-tailed test with rejection on the right. Therefore the p-value is 0.0009, and the null hypothesis is rejected even at an α of 1%.

8.5 F-Test for Equality of Population Variances

In order to test equality of population variances, it is necessary to use the F-distribution named after English statistician Sir Ronald A. ***F***isher. The random variable F is the ratio between the variances of two independent random samples taken from the same normally distributed population (or two different normal populations with the same variance). If the first sample has a size of n_1 and a variance of s_1^2 and the second has size n_2 and variance s_2^2, the F ratio would be s_1^2/s_2^2. Since the numerator and the denominator of this ratio can have any number of degrees of freedom, the F-distribution is specified by degrees of freedom of the numerator, df_1, as well as of the denominator, df_2. These degrees of freedom will be $(n_1 - 1)$ and $(n_2 - 1)$.

The F-distribution has a mean of $df_2/(df_2\text{-}2)$ for $df_2 > 2$ and a variance of

$$\frac{2df_2^2(df_1+df_2-2)}{df_1(df_2-2)^2(df_2-4)} \quad \text{for } df_2 > 4.$$

The *F* ratio cannot be negative, and has the range zero to infinity. The distribution is skewed to the right. The sheet named "F Distn." in Testing Differences can be used to carry out *F* distribution calculations. This template is similar to the Chi-square template. The spreadsheet functions FDIST and FINV are used in the calculations. The formula `=FDIST(x,df₁,df₂)` will return the area to the *right* of *x*. The function `=FINV(α,df₁,df₂)` will return that value of *x* to the right of which lies an area of α.

In the past when spreadsheets did not exist, statisticians had to use *F* distribution tables for calculations. Since *F* distribution is not symmetric and needs two numbers df_1 and df_2 for specification, preparing tables is quite tedious. Restricting the tables to only the right tail areas halves the tedium, and that is what is done in almost all textbooks. The test statistic *F* is so defined that the rejection always occurs on the right tail so that tables that contain only the right tail areas are adequate. With the advent of spreadsheets, this practice is not necessary. However, to conform to the prevailing practice, we shall also define the *F* statistic so as to have rejection on the right tail.

The sheet named F-Test in Testing Differences is shown in Figure 8.5.1. After the input data are entered in the shaded cells, the *F* statistic and the *p*-value appear in cell G6 and G7. The *F* statistic is defined as the ratio of the larger variance to the smaller so that the rejection always occurs on the right tail. Correspondingly, the formula in cell G6 is

```
=MAX(B5:C5)/MIN(B5:C5).
```

The formula in cell G4 is

```
=IF(B5>C5,B4-1,C4-1).
```

The formula in cell G5 is

```
=IF(B5<C5,B4-1,C4-1).
```

The formula for *p*-value in cell G7 is

```
=2*FDIST(G6,G4,G5).
```

The *p*-value that appears cell G7 is for a 2-tailed test. For the currently entered data in the template the *p*-value is very small. Therefore, we reject the null hypothesis that the two variances are equal.

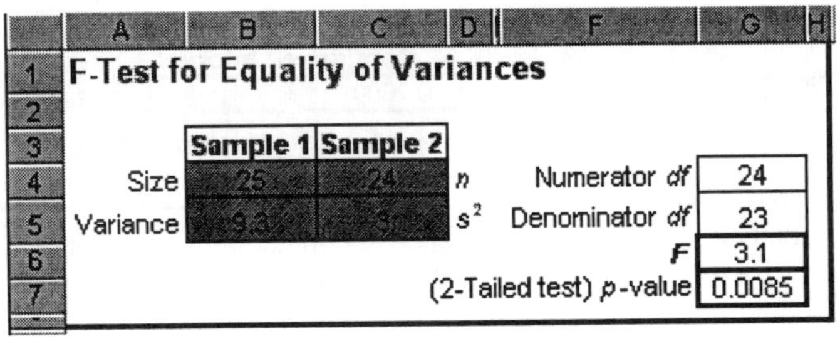

Figure 8.5.1 *F*-Test for Equality of Variances
[F-Test in Testing Differences]

8.6 Exercises

1. Do problem 8-29 from Aczel's book.
2. Do problem 8-50 from Aczel's book.
3. Do problems 8-57, 8-58 and 8-66 from Aczel's book.
4. Do problems 8-61 and 8-62 from Aczel's book.

9. Analysis of Variance

9.1 Introduction

The accuracy with which a marksman can shoot at a target is assessed by asking him to shoot a certain number of times at a target and then measuring the size of the smallest circle that can cover all the bullet marks. Smaller the size, better the accuracy.

Suppose you shoot at a target several times using three different guns. The size of the smallest circle that covers all the marks turns out too large and you are disappointed. On closer examination, you find that the marks made by the three different guns fall into three different circles, as shown in Figure 9.1.1. The triangle joins the centers of the circles. With the pattern as evidence, you suspect that the guns are defective, perhaps in the alignment of their sights, and, in effect, aim at three different places. This defect, you argue, has contributed to the large variance in the marks you made. In other words, you want to refute the "null hypothesis" that the guns are perfect.

But anytime one uses three different guns, their centers are bound to be different. When can we say that the guns are aiming at different spots? Compare the case in Figure 9.1.1 with two others in Figure 9.1.2.

Figure 9.1.1. Marks made by three different guns.

Figure 9.1.2. Two more patterns of marks

In case A, the size of the circles has reduced. In B, the size of the triangle has reduced. It is obvious that case A is a stronger evidence that the guns are aiming at three different spots. Suppose we want a quantitative measure of the strength of the evidence against the null hypothesis. Precisely what quantity in all the cases makes the evidence strong? A meaningful one would be the ratio

$$\frac{\textit{Size of the Triangle}}{\textit{Size of the Circle}}.$$

When this ratio is large enough, we can reject the null hypothesis that the guns are perfect. The size of the triangle measures how far apart the group centers are, and therefore can be termed "Between-Group-Variance." The size of the circle measures the variance within each group and therefore can be termed "Within-Group-Variance." The ratio then becomes

$$\frac{\textit{Between Group Variance}}{\textit{Within Group Variance}}.$$

The concept behind Analysis of Variance (ANOVA) is similar. When a population consisting of many groups shows a large variance, we analyze the variance to see if the different groups have different centers (or means) and thus contribute to the variance. The hypotheses are formulated as

H_0: All group means are equal

H_1: Not all group means are equal.

Under some not too restrictive assumptions the ratio of Between-Group-Variance to Within-Group-Variance follows an *F*-distribution. Thus, it becomes a 1-tailed *F*-test with rejection on the right.

9.2 Single Factor ANOVA

We shall see how Between-Group-Variance and Within-Group-Variance are calculated with numerical data (rather than bullet marks). Recall that the calculation of variance requires taking the deviations from the mean, squaring the deviations, summing the squared deviations and then dividing the sum of squared deviations (SSD) by the degrees of freedom. Figure 9.2.1 shows the sheet named Concepts in the ANOVA workbook, where the necessary calculations have been made. The data is from Exercise 9-18 of Aczel, where the flight ranges of three different aircraft designs A, B and C are compared to test if the different models have different means. The null hypothesis is that all group means are equal.

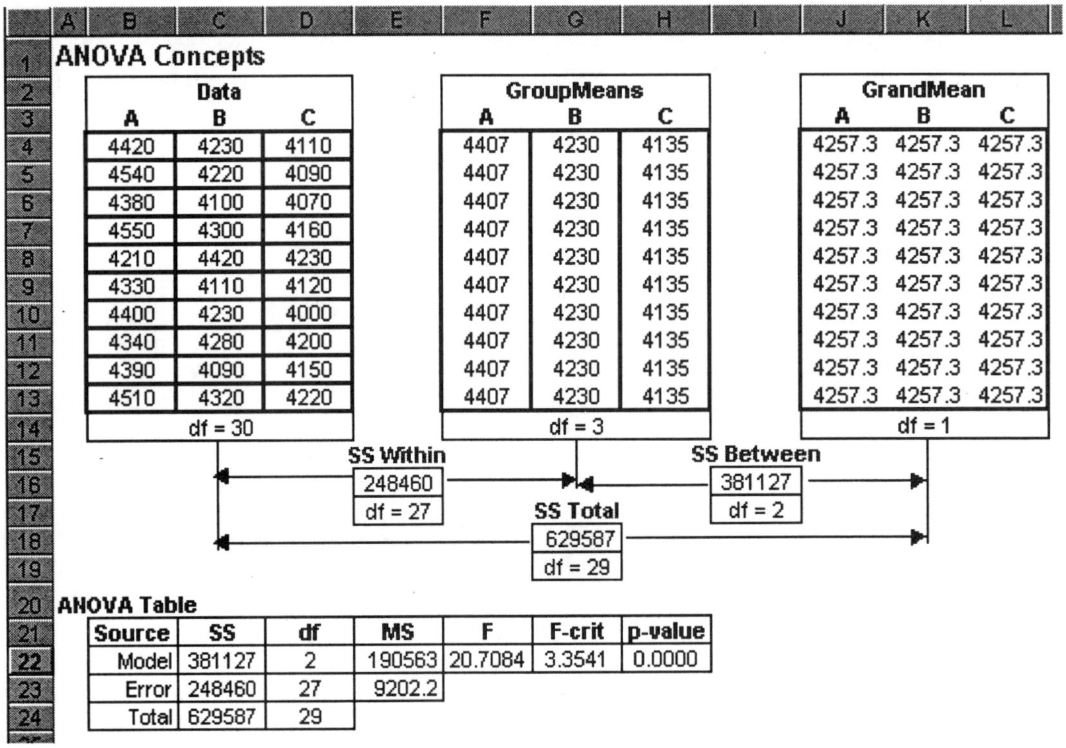

Data A	Data B	Data C	GroupMeans A	GroupMeans B	GroupMeans C	GrandMean A	GrandMean B	GrandMean C
4420	4230	4110	4407	4230	4135	4257.3	4257.3	4257.3
4540	4220	4090	4407	4230	4135	4257.3	4257.3	4257.3
4380	4100	4070	4407	4230	4135	4257.3	4257.3	4257.3
4550	4300	4160	4407	4230	4135	4257.3	4257.3	4257.3
4210	4420	4230	4407	4230	4135	4257.3	4257.3	4257.3
4330	4110	4120	4407	4230	4135	4257.3	4257.3	4257.3
4400	4230	4000	4407	4230	4135	4257.3	4257.3	4257.3
4340	4280	4200	4407	4230	4135	4257.3	4257.3	4257.3
4390	4090	4150	4407	4230	4135	4257.3	4257.3	4257.3
4510	4320	4220	4407	4230	4135	4257.3	4257.3	4257.3
df = 30			df = 3			df = 1		

ANOVA Table

Source	SS	df	MS	F	F-crit	p-value
Model	381127	2	190563	20.7084	3.3541	0.0000
Error	248460	27	9202.2			
Total	629587	29				

Figure 9.2.1 One Way ANOVA Concepts
[Concepts in ANOVA]

The calculation process requires the construction of the Data matrix, the GroupMeans matrix and the GrandMean matrix. The grand mean is the mean of all data, which, in the bullet marks analogy, corresponds to the center of the triangle. We shall see the construction of the spreadsheet step by step.

<u>To construct the Data matrix</u>

- Enter the data in the range B4:D13
- Apply heavy outline border to cell B4
- Click on the format painter icon
- Drag the mouse over the range B4:D13 (This creates the desired border pattern.)
- Name the range B4:D13 as `Data`

<u>To construct the GroupMeans matrix</u>

- Select the range F4:F13
- Enter the formula `=AVERAGE(B4:B13)`
- Press `CTRL+SHFT+Enter`
- Apply heavy outline border

- Drag the AutoFill button from cell F13 to cell H13.
- Name the range F4:H13 as `GroupMeans`

<u>To construct the GrandMean matrix</u>

- Select the range J4:L13
- Enter the formula `=AVERAGE(B4:D13)`
- Press `CTRL+SHFT+Enter`
- Apply heavy outline border
- Name the range J4:L13 as `GrandMean`

Next let us consider Within-Group-Variance. For this we need the Sum of Squared (SS) deviations within groups. To calculate this SS we must take the deviation of each data point from its corresponding group mean, square the deviation, and sum the squared deviations. The actual calculation will be

$$(4420 - 4407)^2 + (4540 - 4407)^2 + \ldots + (4220 - 4135)^2$$

The function SUMXMY2 can be used to calculate this quantity in one stroke. SUMXMY2 stands for the SUM of (X Minus Y)^2. In cell E16, the formula

`=SUMXMY2(Data,GroupMeans)`

has been entered. This calculates SS_{within} as 248460. This SS must be divided by its degrees of freedom (*df*) to get the variance. Since this SS is arrived at by taking the difference between Data and GroupMeans, its *df* is the difference between the *df*'s of Data and GroupMeans matrices. To determine the *df* of the Data matrix, we ask the question, how many *different* values are possible in that matrix. There are 30 numbers and all of them can be different. Hence the matrix has $df = 30$. The GroupMeans can have only 3 *different* values corresponding to the three groups, and hence its $df = 3$. The *df* of SS_{within} is therefore $30 - 3 = 27$. Dividing SS_{within} by its degrees of freedom we get the Within-Group-Variance as 248460/27 = 9202.2. The symbol used for Within-Group-Variance is MS_{within} where MS stands for Mean Square.

To calculate Between-Group-Variance, we take the deviation of the group means from the grand mean, square them and add the squares. In cell I16 the formula

`=SUMXMY2(GroupMeans,GrandMean)`

has been entered. This calculates $SS_{between}$ as 381127. Its *df* is the difference between those of GroupMeans and GrandMean matrices which works out to $3 - 1 = 2$. The Between-Group-Variance is now calculated as 381127/2 = 190563. The symbol for this is $MS_{between}$.

Let us make the following assumptions.

A1. All the groups (populations) are normally distributed.

A2. All the group variances are equal.

A3. The samples from the groups are random and independent.

Among these assumptions, **A1** is somewhat restrictive, **A2** is the most restrictive and **A3** is the least restrictive. Going to the bullet marks analogy, **A2** would mean that all the circles are of the same size. This assumption has been tacitly made in the diagrams. When the circles differ in size, it is intuitively clear that the formulas and calculations would become messy. Hence the assumption. The setup is pictured in Figure 9.2.2.

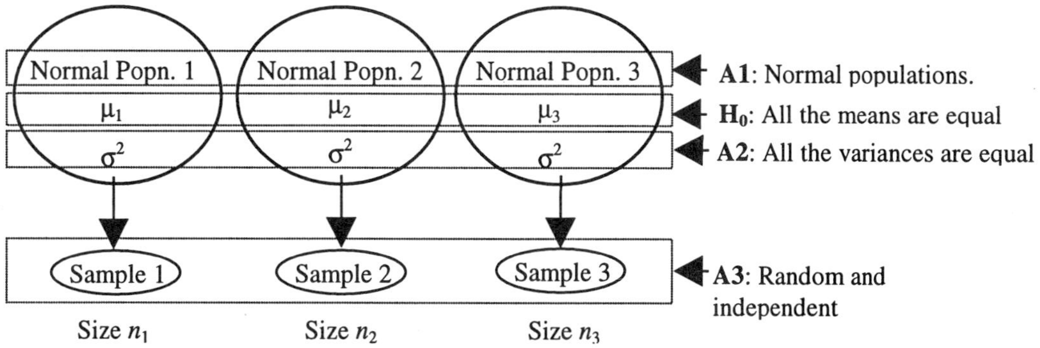

Figure 9.2.2. The Setup for ANOVA

Sir Ronald Fisher proved that under the assumptions **A1-A3**, if the null hypothesis (that all group means are equal) is true, then both MS_{within} and $MS_{between}$ are unbiased estimates of the same variance σ^2 (of the groups). Hence their ratio should follow an *F*-distribution.

The ratio works out to 190563/9202.2 = 20.7084. The numerator *df* is 2 and the denominator *df* is 27. The *p*-value calculated using the FDIST function is almost zero. Hence the evidence is extremely significant and the null hypothesis is rejected. This establishes the alternative hypothesis that different models have different means. Looking at the group means, one can add that Model A has the largest range and Model C the smallest.

These calculations have been organized in the form of the ANOVA table in the range B21:H24. In the first column marked Source, the entries are Model, Error and Total . These may be interpreted as follows: The total variance in the data is proportional to the Total SS of 629587 which may be split into 381127 due to differences between the models ($SS_{between}$) and 248460 due to other unknown random causes, or errors (SS_{within}). Note that SS_{total} calculated in cell G18 using the formula

`=SUMXMY2(Data,GrandMean)`

is also the sum of $SS_{between}$ and SS_{within}. This proves the interesting relationship

$$SS_{total} = SS_{between} + SS_{within}$$

The calculation of the SS values in cells E16, I16 and G18 are only for illustration purposes. While solving a problem, these values are calculated directly in the ANOVA table. The ANOVA table formulas are:

cell C22: `=SUMXMY2(GroupMeans,GrandMean)`
cell C23: `=SUMXMY2(Data,GroupMeans)`
cell C24: `=SUMXMY2(Data,GrandMean).`
cell F22: `=E22/E23`
cell G22: `=FINV(5%,D22,D23)`
cell H22: `=FDIST(F22,D22,D23).`

The three matrices can also be given another interpretation. Suppose the null hypothesis that all group means are equal is true. Suppose further that there were no errors. Then all the observations should have been uniformly equal to the grand mean. Under this ideal condition, the data would have looked like the GrandMean matrix. Similarly, if the alternative hypothesis is true and there were no errors, then the data would have looked like the GroupMeans matrix. In this sense, the GrandMean matrix and the GroupMeans matrix represent the null and alternative hypotheses respectively.

To solve another problem using this same template, one should construct the two matrices Data and GroupMeans and name them properly. Move the ANOVA table around to make space, if necessary.

Note 1: Although the calculations seem complex with numerous intermediate quantities to be calculated, the array formula and the AutoFill button features of Excel make it impressively easy. One further simplification possible is the use of DEVSQ function. The DEVSQ function automatically calculates the (grand) mean and finds the sum of squared deviations from the mean. In other words, `=SUMXMY2(Data,GrandMean)` and `=DEVSQ(Data)` would return the same value, namely, SS_{total}. Similarly, the formula `=DEVSQ(GroupMeans)` would return. The difference between SS_{total} and $SS_{between}$ would yield SS_{within}. The use of DEVSQ thus avoids the need for the GrandMean matrix. The sheet marked "DEVSQ" has the same problem solved using DEVSQ without the GrandMean matrix. After the data is entered, this method should take only a minute or two to solve the problem. The author recommends this method over the Analysis ToolPak macro described later.

Note 2: A great advantage of solving a problem on a spreadsheet as suggested above is that when any change is made in the data, the calculated results update immediately. A disadvantage of the method is that if the sample sizes are not equal, the method becomes less efficient because the AutoFill button cannot be used to copy the array formulas.

Suppose, in the above problem, we drop the last two observations in Model B, and the last one in Model C. Then the sample sizes are 10, 8 and 9. This modified problem has been solved in the sheet named "Unequal". The sheet is seen in Figure 9.2.3. Here, after constructing the first column of GroupMeans matrix in the range F4:F13, the AutoFill button cannot be used to copy it to the right. The other two columns must be constructed individually as follows.

- Select the range G4:G11
- Enter the formula `=AVERAGE(C4:C11)`

- Press `CTRL+SHFT+Enter`
- Select the range H4:H12
- Enter the formula `=AVERAGE(D4:D12)`
- Press `CTRL+SHFT+Enter`

One needs to be careful in calculating the *df*. The *df* for Data matrix is only 27 now because there are only 27 numbers in the matrix and all of them can be different. In cell C14, the formula `=COUNT(B4:D13)` has been used. The *df* for the GroupMeans matrix is still 3 since there can be three different numbers in the matrix.

ANOVA with Unequal Observations

Data		
A	B	C
4420	4230	4110
4540	4220	4090
4380	4100	4070
4550	4300	4160
4210	4420	4230
4330	4110	4120
4400	4230	4000
4340	4280	4200
4390		4150
4510		
df = 27		

GroupMeans		
A	B	C
4407	4236.3	4125.6
4407	4236.3	4125.6
4407	4236.3	4125.6
4407	4236.3	4125.6
4407	4236.3	4125.6
4407	4236.3	4125.6
4407	4236.3	4125.6
4407	4236.3	4125.6
4407		4125.6
4407		
df = 3		

ANOVA Table

Source	SS	df	MS	F	F-crit	p-value
Model	383099	2	191549	21.6420	3.4028	0.0000
Error	212420	24	8850.8			
Total	595519	26				

Figure 9.2.3 Unequal sample sizes
[Unequal in ANOVA]

The *p*-value is almost zero and therefore the null hypothesis is rejected. Model A has the largest range and Model C has the smallest.

Although this method has lost some elegance in going from the equal sample sizes case to the unequal case, the author still recommends this method over the Analysis ToolPak method described below, especially to students learning the technique for the first time. The ToolPak method is suitable if one has mastered the ANOVA technique.

Refer to Figure 9.2.5 which shows the sheet named "ToolPak" in the ANOVA workbook. After the data is entered in range D4:D13, select the **Data Analysis...** command under the **Tools** menu. In the dialog box that appears, choose **Anova: Single Factor**. Another dialog box shown in Figure 9.2.4 appears.

In this dialog box, the **Input Range** is filled in as `B3:D13` to include the labels in the top row (Row number 3). See Figure 9.2.5. The groups are arranged in **Columns** with **Labels in First Row**. The **Alpha** we wish to use is 5% (for calculating *F*-Critical). We want the output on the same sheet starting at cell F2. After making these entries, click the **OK** button. The ANOVA calculations are carried out and the output appears as seen in Figure 9.2.5.

In the output, under SUMMARY, the sample statistics Count, Sum, Average and Variance appear. Under ANOVA, the standard ANOVA table appears.

The greatest advantage of this method is its speed. After the data is entered, it takes only a few seconds to solve a problem. The greatest disadvantage is that the results will not update if the data is changed. The macro should be used once again to update.

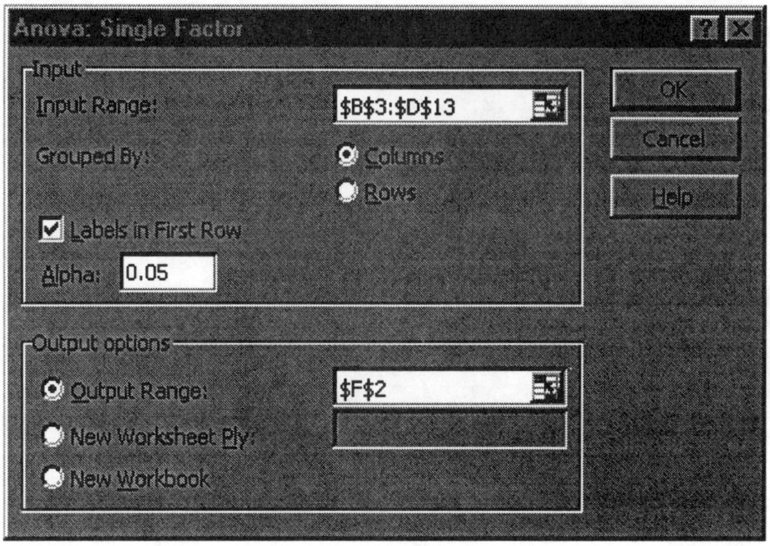

Figure 9.2.4 Single Factor ANOVA Dialog Box

	A	B	C	D	E	F	G	H	I	J	K	L
1	Solved Using Analysis ToolPak											
2		Data				Anova: Single Factor						
3		A	B	C								
4		4420	4230	4110		SUMMARY						
5		4540	4220	4090		*Groups*	*Count*	*Sum*	*Average*	*Variance*		
6		4380	4100	4070		A	10	44070	4407	11023.3		
7		4550	4300	4160		B	10	42300	4230	11400		
8		4210	4420	4230		C	10	41350	4135	5183.33		
9		4330	4110	4120								
10		4400	4230	4000								
11		4340	4280	4200		ANOVA						
12		4390	4090	4150		*Source of*	*SS*	*df*	*MS*	*F*	*P-value*	*F crit*
13		4510	4320	4220		Between (	381127	2	190563	20.7084	3.5E-06	3.35413
14						Within Grc	248460	27	9202.22			
15												
16						Total	629587	29				

Figure 9.2.5 Analysis ToolPak Method for Single Factor ANOVA
[ToolPak in ANOVA]

9.3 Two Factor ANOVA

Often, a random variable is affected by more than one factor and we may want to analyze how each factor contributes to the variance of the random variable. This is the case that we shall see in this section. Although the same principle can be extended to more than two factors, we shall stop with two.

Suppose the brightness of clothes after a wash depends on both the washing machine and the detergent used. We wish to analyze the effect of each of these two factors on the brightness. We wash several clothes using different combinations of machine and detergent and measure their brightness on a scale of 1 to 25. Larger number signifies higher brightness. Let the data be as shown in Figure 9.3.1.

Before we proceed further, we shall distinguish between two different cases of selecting the machines and the detergents. If there are numerous brands of machines and detergents and we select the machines and the detergents at random from the available brands, then it would be a *Random-Effects Model.* The results we obtain about the effects of machines and detergents would then apply to all the machines and all the detergents. If we were interested in only particular brands of machines and detergents and therefore selected only those, then it would be a *Fixed-Effects Model* and our results would apply to only those brands of machines and detergents.

As the data shows three different detergents and four different machines have been used. Further, for each combination of machine and detergent, three *replications* have been made. In the Data matrix, every combination of machine and detergent is called a *cell*, and there are 3 x 4 = 12 cells. In each cell there are 3 replications, making a total of 12 x 3 = 36 observations.

Unlike the single factor case, the two factor case includes what is known as an *interaction effect.* A machine by itself has some effect on brightness and a detergent does similarly. In addition, there could be an effect due to the particular combination of machine and detergent. A machine may work better with a particular detergent and thus can have a positive interaction effect. Negative interactions are also possible. Since the detergents are along the rows, they are called the row factor; the machines, being along the columns, are called the column factor.

There are three null hypotheses corresponding to the three effects. Each one simply states that the effect is zero. If a null hypothesis is rejected, it means that the effect is not zero, i.e., it is significant.

We make the following assumptions similar to those of the single factor case.

A1. The model is

Observed value = Grand Mean + Row Factor Effect + Column Factor Effect + Interaction Effect + Error.

In other words, the effects are additive.

A2. The error in the above model is normally distributed with zero mean and constant variance.

A3. The samples are random and independent.

Under these assumptions, if the null hypotheses are true, then the ratio obtained by dividing the MS due to an effect by the MS due to error will follow an *F*-distribution. We can therefore apply *F*-tests for all three effects as in the single factor case.

Two Factor ANOVA

	Data				RowMeans			
	M1	M2	M3	M4				
D1	17	17	11	12	16	16	16	16
	18	16	16	18	16	16	16	16
	16	20	20	16	16	16	16	16
D2	10	13	12	7	10	10	10	10
	8	12	8	11	10	10	10	10
	6	11	17	9	10	10	10	10
D3	18	16	15	17	15	15	15	15
	14	15	13	19	15	15	15	15
	11	6	20	18	15	15	15	15
	df = 36				df = 3			

ANOVA Table

Source	SS	df	MS	F	F-Crit	*p*-Value
Detergent	247.72	2	123.86	12.2837	3.4028	0.0002
Machine	11.194	3	3.7315	0.3701	3.0088	0.7753
Interaction	94.056	6	15.676	1.5546	2.5082	0.2036
Error	242	24	10.083			
Total	594.97	35				

Conclusions:

Detergent has significant effect

Machine does not have significant effect

Interaction does not have significant effect

ColMeans				CellMeans			
13	14	15	14	17	18	16	15
13	14	15	14	17	18	16	15
13	14	15	14	17	18	16	15
13	14	15	14	8	12	12	9
13	14	15	14	8	12	12	9
13	14	15	14	8	12	12	9
13	14	15	14	14	12	16	18
13	14	15	14	14	12	16	18
13	14	15	14	14	12	16	18
df = 4				df = 12			

Figure 9.3.1 Two Factor ANOVA Calculations
[2 Factor in ANOVA]

Figure 9.3.1 shows the sheet named "2 Factor" in the ANOVA workbook. As seen in this figure, four matrices are needed for the calculation. After the Data matrix is entered and named as `Data`, the remaining matrices can be constructed as follows.

To construct the <u>RowMeans</u> matrix:

- Select the range G4:J6
- Enter the formula `=AVERAGE(B4:E6)`
- Press `CTRL+SHFT+Enter`
- Apply heavy outline border
- Drag the AutoFill button from J6 to J12
- Name the range G4:J12 `RowMeans`

To construct the <u>ColMeans</u> matrix:

- Select the range B16:B24
- Enter the formula `=AVERAGE(B4:B12)`
- Press `CTRL+SHFT+Enter`
- Apply heavy outline border
- Drag the AutoFill button from B24 to E24
- Name the range B16:E24 as `ColMeans`

To construct the <u>CellMeans</u> matrix:

- Select the range G16:G18
- Enter the formula `=AVERAGE(B4:B6)`
- Press `CTRL+SHFT+Enter`
- Apply heavy outline border
- Drag the AutoFill button from G18 to J18
- Drag the AutoFill button again from J18 to J24
- Name the range G16:J24 as `CellMeans`

The SS values are now calculated using the DEVSQ function. The important thing to note here is how $SS_{CellMeans}$ and its *df* are related to $SS_{Interaction}$ and its *df*. The relationships are:

$$SS_{CellMeans} = SS_{RowMeans} + SS_{ColMeans} + SS_{Interaction}$$

$$df(SS_{CellMeans}) = df(SS_{RowMeans}) + df(SS_{ColMeans}) + df(SS_{Interaction})$$

These relationships are used in calculating $SS_{Interaction}$ and its *df*. Specifically, DEVSQ(CellMeans) will return $SS_{CellMeans}$. Subtracting $SS_{RowMeans}$ and $SS_{ColMeans}$ from this will yield $SS_{Interaction}$. The *df* for Interaction is similarly calculated. The SS and *df* calculations are shown schematically in the diagram below.

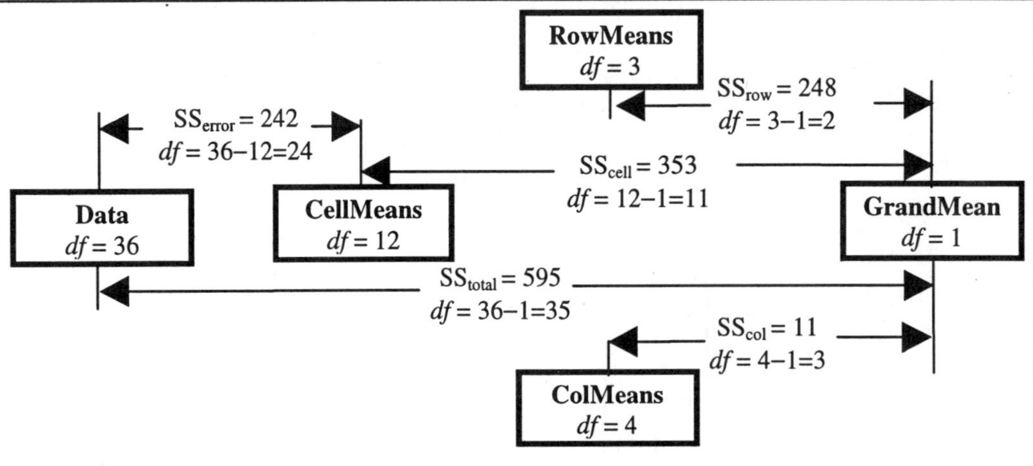

$$\begin{aligned} SS_{interaction} &= SS_{cell} - SS_{row} - SS_{col} \\ &= 353 - 248 - 11 \\ &= 94 \\ df(SS_{interaction}) &= df(SS_{cell}) - df(SS_{row}) - df(SS_{col}) \\ &= 11 - 2 - 3 \\ &= 6 \end{aligned}$$

Figure 9.3.2 Schematics of *df* and SS Calculations in 2 Factor ANOVA

On the template, important formulas used for constructing the ANOVA table are:
cell M4: `=DEVSQ(RowMeans)`
cell M5: `=DEVSQ(ColMeans)`
cell M6: `=DEVSQ(CellMeans)-M4-M5`
cell M7: `=M8-M4-M5-M6` [`=SUMXMY2(Data,CellMeans)` can also be used]
cell M8: `=DEVSQ(Data)`
cell O4: `=M4/N4` [Copied down to O7]
cell P4: `=O4/$O$7` [Copied down to P6]
cell Q4: `=FINV(5%,N4,$N$7)` [Copied down to Q6]
cell R4: `=FDIST(P4,N4,$N$7)` [Copied down to R6]

The p-value for Detergent Effect is 0.0002 and therefore we declare that Detergent has significant effect on brightness. In other words, we reject the null hypothesis that the average brightness is the same for all detergents. The *p*-values of Machine and Interaction Effects are large. Hence we cannot reject the null hypotheses regarding them.

To use the template to solve another problem, one should construct the four matrices Data, RowMeans, ColMeans and CellMeans and name them properly. In addition, the number of replications per cell should also be specified. It is needed to determine the *df*'s.

<u>Note</u>: It is possible to use Analysis ToolPak to carry out a 2-Factor ANOVA. This has been done in the sheet named "2 Factor ANOVA (2)" in Chapter9.xlt. See Figure 9.3.3. The steps involved are:

- Under the **Tools** menu select the **Data Analysis...** command
- In the dialog box that appears, select **Anova: Two-Factor with Replication**
- In the next dialog box that appears, in the **Input Range Box** fill in `A3:E12`. [Note: You must include the labels for the rows and the columns.]

	A	B	C	D	E	F	G	H	I	J	K	L
1	Two Factor ANOVA Solved Using Analysis ToolPak											
2		Data					Anova: Two-Factor With Replication					
3		M1	M2	M3	M4							
4	D1	17	17	11	12		SUMMARY	M1	M2	M3	M4	Total
5		18	16	16	18		D1					
6		16	20	20	16		Count	3	3	3	3	12
7	D2	10	13	12	7		Sum	51	53	47	46	197
8		8	12	8	11		Average	17	17.6667	15.6667	15.3333	16.4167
9		6	11	17	9		Variance	1	4.33333	20.3333	9.33333	7.35606
10	D3	18	16	15	17							
11		14	15	13	19		D2					
12		11	6	20	18		Count	3	3	3	3	12
13							Sum	24	36	37	27	124
14							Average	8	12	12.3333	9	10.3333
15							Variance	4	1	20.3333	4	9.15152

Figure 9.3.3 2-Factor ANOVA Solved Using Analysis ToolPak
[2 Factor ToolPak in ANOVA]

- In the **Rows per sample** box, enter `3` [This is the number of replications in each cell.]
- Leave the **Alpha** as `0.05`
- Click on **Output Range** and enter in the box next to it, `G2`
- Click the **OK** button

Excel carries out all the calculations and displays the output starting in cell G2. In Figure 9.3.3, only part of the output is shown. At the bottom of the output, the ANOVA table appears.

The author discourages the use of the Analysis ToolPak for 2-Factor ANOVA because the results will not update when a change is made in the data.

9.4 Block Designs

9.4.1 Randomized Complete Block Design

If the interaction effect is to be tested, there should be at least 2 replications in each cell. Otherwise, the *df* for error will go to zero. If it is known that no interaction effect exists, then one may use designs with only one replication per cell. A particular case with only one replication per cell is the Randomized Block Design.

Suppose the brightness of clothes washed in a machine depends on the color of the clothe besides the machine's characteristics. To test 3 different machines, we may then form "blocks" of clothes. We may select 3 clothes that have the same color, call it a block, and assign them randomly to the 3 machines. We then assign many (8 in Figure 9.4.1) such blocks to each machine. The result is a *Randomized Complete Block Design.*

Figure 9.4.1 shows the sheet named "Rand. Comp. Block". In this template, a randomized block design problem solved using three matrices Data, RowMeans and ColMeans. The only difference here is that there is no Interaction effect. Since the *p*-value for the Machine effect is less than 1%, we reject the null hypothesis and declare that Machines have significant effect.

Randomized Complete Block Design

	Data			RowMeans			ColMeans			ANOVA Table						
	M1	M2	M3	M1	M2	M3	M1	M2	M3	Source	SS	df	MS	F	F-Crit	*p*-Value
B1	25	22	18	22	22	22	19	18	13	Block	213.3	7	30.48	2.6848	2.7642	0.0549
B2	20	23	15	19	19	19	19	18	13	Machine	149.1	2	74.54	6.5669	3.7389	0.0097
B3	23	19	10	17	17	17	19	18	13	Error	158.9	14	11.35			
B4	19	18	11	16	16	16	19	18	13	Total	521.3	23				
B5	12	16	19	16	16	16	19	18	13							
B6	13	11	9	11	11	11	19	18	13							
B7	23	19	11	18	18	18	19	18	13							
B8	18	13	13	15	15	15	19	18	13							
	df = 24			df = 8			df = 3									

Figure 9.4.1. Randomized Complete Block Design
[Rand. Comp. Block in ANOVA]

It is possible to solve Randomized Complete Block Design problems using Analysis ToolPak. The sheet named "Rand. Comp. Block (2)" solves the same problem above using Analysis ToolPak.

9.4.2 Repeated Measures Design

A special case of Randomized Block Design is the Repeated Measures Design where the measurements along a row are repeated on the same subject. Suppose we wish to measure the appeal of four different package designs on consumers. We may ask several potential customers to rate the package designs on a scale of 1 to 10. In the Data matrix, each row will correspond to one and the same subject rather than four different subjects forming a block. Errors will be further reduced because of this.

The calculation process is the same as Randomized Complete Block Design and therefore there is no separate template for this case.

9.5 Exercises

1. Do exercise 9-19 in Aczel's book.
2. Do exercise 9-42 in Aczel's book.
3. Do exercise 9-62 in Aczel's book.
4. Do exercise 9-64 in Aczel's book.

10. Simple Linear Regression and Correlation

10.1 Introduction

A pair of random variables, such as advertising expenses and sales of a company, may be related to each other. If so, it would be useful to quantify their relationship so that we can predict one of them knowing the value of the other, or control one of them by controlling the other. The simplest kind of relationship between two variables X and Y is the linear relationship where the graph of the relationship is a straight-line as shown in Figure 10.1.1.

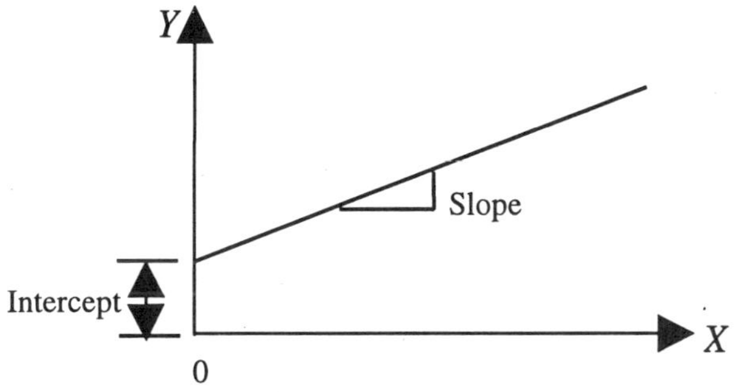

Figure 10.1.1. A Simple Linear Relationship

This relationship can be written as an equation

$$Y = \text{Intercept} + \text{Slope}*X$$

so that once the two quantities, Intercept and Slope, are known the entire relationship is known. The objective in a regression analysis is to find relationships like this between two or more variables. When there are only two variables and the relationship is linear we call the analysis a *simple linear regression.*

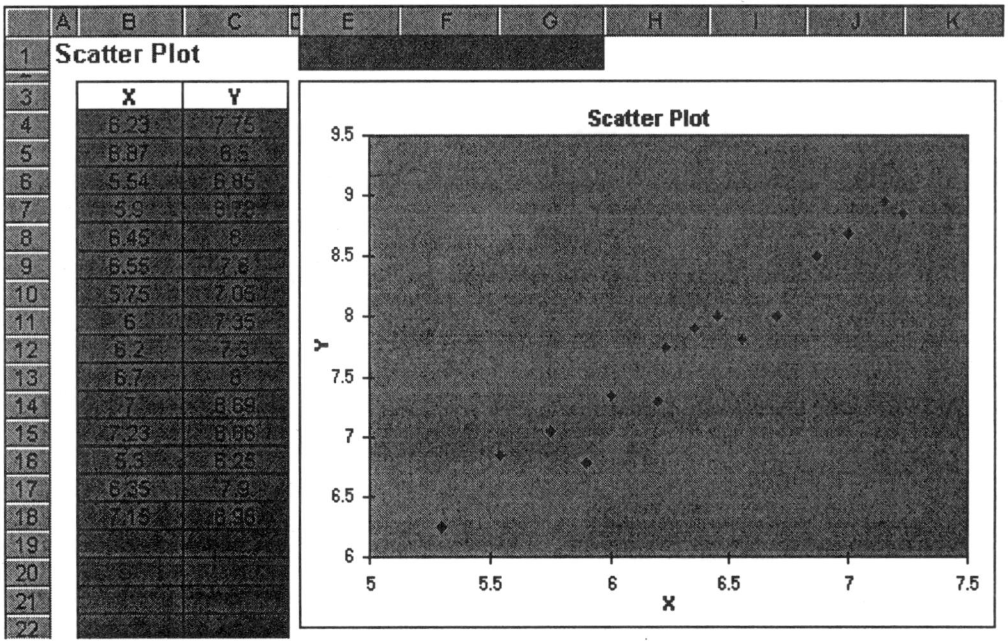

Figure 10.1.2. Regression Data and Scatter Plot
[Scatter Plot in Simple Regression]

Consider the case where the two variables are the advertising expense and the sales of a business organization. It is reasonable to expect that a change in advertising expense will cause a change in sales. Thus sales depends upon advertising expense and we call sales the *dependent variable* and advertising expense the *independent variable*. We use the symbol Y for the dependent variable and X for the independent variable.

Suppose we suspect that a linear relationship exists between X and Y and wish to quantify the relationship in the form of an equation. We collect sample data about the two variables and plot them on a graph using the template shown in Figure 10.1.2. The graph is called a *Scatter Plot* since it shows a scatter of points. If the axes of the plot are not properly scaled. then re-scale the axes. To re-scale the axes,

- Click on the X-axis
- Choose the **Selected Axis...** command under the **Format** menu
- Select the **Scale** tab
- In the dialog box enter the desired **Minimum** and **Maximum** and click the **OK** button
- Click on the Y-axis and do the same.

The scatter plot suggests that there is a relationship between the two variables because the points seem to fall along a straight-line. We further note that as X increases, Y also increases. In this case, we say that there is a *positive correlation* between the two variables. If the straight-line has a negative slope, then as one variable increases the other decreases, and we would then say there is a *negative correlation*. If the points are scattered all over the graph or if they suggest a horizontal straight-line, then we would say that there is no correlation between the variables. See Figure 10.1.3.

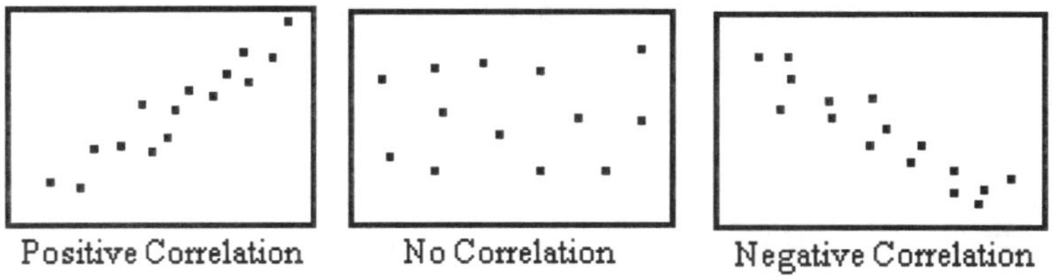

Figure 10.1.3. Types of Correlations

When a positive or a negative correlation is suggested by the scatter plot, one is persuaded to do further calculations and quantify the relationship. If no correlation is suggested, then the regression analysis might not be fruitful. Our case of X and Y does suggest a positive correlation and hence we proceed to quantify the relationship by fitting a straight-line that best fits the scatter of points. This straight-line is called the *regression line*.

Different people may fit different straight-lines to the same data. The natural question is how do we define the line of *best fit*. When a point in the scatter plot does not fall right on the regression line there is an error. Let us define error as

Error = Actual Y value − Y value predicted by the regression line.

The actual Y value is denoted by y, the predicted value by $\hat{y}$ and the error by e. Thus

$$e = y - \hat{y}.$$

See Figure 10.1.4. Note that the error is positive when the data point is above the regression line and negative when it is below.

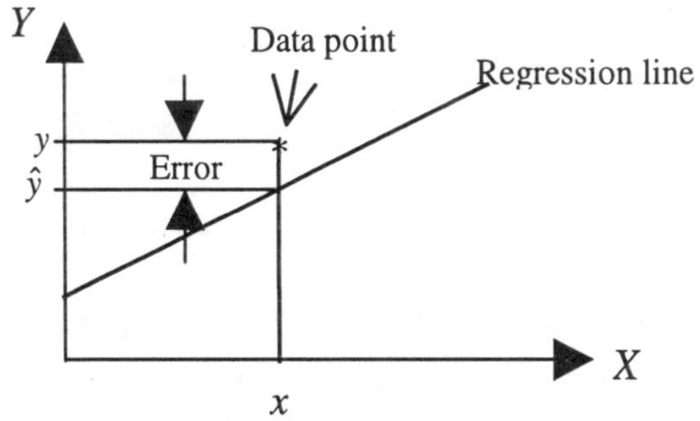

Figure 10.1.4 Error

Intuitively, we can say that the line of best fit is the one that minimizes the "overall error." What would be a good measure of the overall error? It cannot be the sum of all the errors because the negative and positive errors will mostly cancel each other and we will get zero or close to zero as the sum, no matter how big the errors are. What are the alternatives? There are three popular alternatives here. Instead of minimizing the sum of errors, we may minimize

1. the sum of the *squares* of all the errors (SSE)
2. the sum of the *absolute values* of all errors
3. the *median* of the *squares* of all the errors.

The first alternative of minimizing SSE has the following advantages:

i. It is amenable to simple optimization techniques. The Solver tool in Excel is capable of doing it. Also, one can use calculus. [The use of calculus leads to the normal equations described later.]

ii. It produces a unique regression line.

iii. If we make the following assumptions **A1-A4**, *t* and *F* tests are possible to test hypotheses regarding the regression.

A1. $Y = \beta_0 + \beta_1 X + \varepsilon$

where β_0 denotes the intercept, β_1 the slope and ε the error.

A2. The error ε is normally distributed with zero mean and finite variance σ^2. The variance σ^2 is constant for all values of X. Further, the errors at one value of X is not correlated with those at any other value of X.

A3. The error is not correlated with X.

A4. The X values of the sample are fixed and not random

iv. Under the assumptions **A1-A4**, the intercept and the slope calculated from sample data, denoted by b_0 and b_1, are *unbiased* estimates of the true parameters β_0 and β_1.

The disadvantages of this alternative are:

i. SSE has no practical significance. If we are not interested in any hypothesis tests, minimizing SSE does not have any practical advantage.

ii. The squaring of errors to calculate SSE disproportionately amplifies large errors relative to small errors. As a result, a very large error caused by an outlier data point will adversely affect the regression line. In Figure 10.1.5, one outlier point has distorted the regression line, making it non-representative of the true relationship.

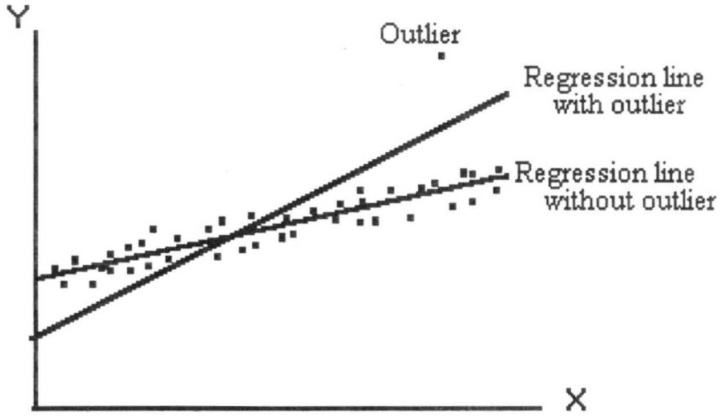

Figure 10.1.5 Adverse Effect of an Outlier in Min SSE Method

The second alternative of minimizing the sum of absolute errors has the following advantages:

i. No distortion of squaring the errors takes place.

ii. Not affected much by outliers.

The disadvantages are:

i. The solution may not be unique.

ii. The calculations are not easy.

iii. Hypothesis tests are not possible.

The third alternative of minimizing the median of squared errors has the following advantage:
i. Not affected by outliers at all. For this reason this method is called *robust regression.*
The disadvantages are:
i. The calculations are difficult.
ii. Hypothesis tests are not possible.
iii. The solution may not be unique, although this possibility is rare.

The advantages of the first alternative of minimizing SSE are quite appealing and therefore this method is by far the most popular approach to regression. Because the method finds the least sum of squared errors, it is called the *Least Squares Method.* In the rest of this chapter, we will consider only this method.

The spreadsheet offers as many as seven different ways of carrying out the least squares method. These are listed below. We shall consider them one-by-one.

1. Visual method
2. Solver method
3. LINEST method
4. Matrix method
5. Analysis ToolPak method
6. Normal Equations method
7. Trendline Method

10.2 The Visual Method

10.2.1 The Method

In the visual method, one visually adjusts the regression line until the SSE is sufficiently minimized. The sheet named "Visual" in Simple Regression is shown in Figure 10.2.1. In this template, a regression line has been superposed on the scatter plot. If the data for *X* and *Y* are changed, the axes may have to be re-scaled to bring all the points and the regression line into view. Re-scaling the axes is explained in the previous section.

When one changes the intercept or the slope in cells F5 and G5, the regression line changes on the graph accordingly. The sum of the squared errors, SSE, is also updated. By watching the graph and the SSE, one can adjust the intercept and slope until one finds the line that minimizes SSE. While adjusting slope and intercept, one has to jockey back and forth. This means that if you have found the best slope and then adjust the intercept, the slope may not be the best anymore. You will have to adjust it again. And then the intercept may not be the best ... Thus, it is a long trial and error process. The process is halted when SSE cannot be decreased appreciably anymore.

10.2.2 Advantages and Disadvantages of the Method

The advantages of the method are:
1. The user has a good picture of what is going on. This is especially good for one who is learning the technique for the first time.
2. This method is useful if there are some unusual constraints. For instance, if we had a constraint that the regression line can have at most five observations lying above it (i.e., at most five positive errors), then it might not be possible to do it any other way.

The disadvantages of the method are:
1. It is a tedious trial and error process.
2. If the data is changed, the trial and process must be carried out again.
3. The intercept and slope obtained by this method will not be as accurate as in other methods.
4. Many other quantities will have to be hand calculated for further analyses such as hypothesis tests.

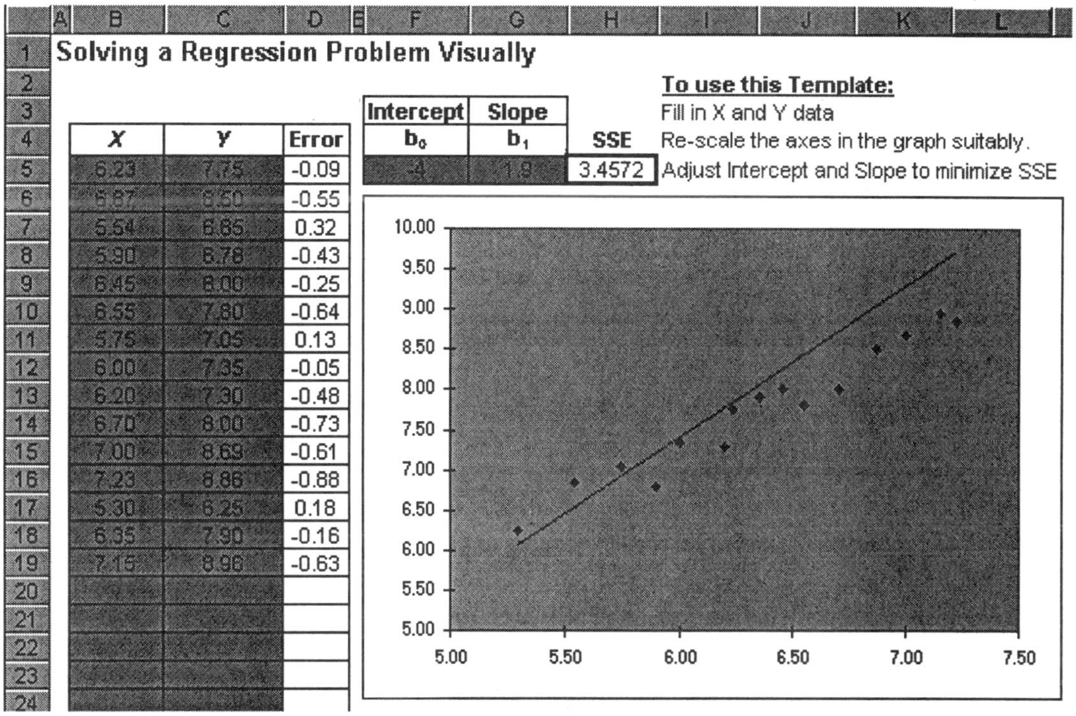

Solving a Regression Problem Visually

X	Y	Error
6.23	7.75	-0.09
6.87	8.50	-0.55
5.54	6.85	0.32
5.90	6.78	-0.43
6.45	8.00	-0.25
6.55	7.80	-0.64
5.75	7.05	0.13
6.00	7.35	-0.05
6.20	7.30	-0.48
6.70	8.00	-0.73
7.00	8.69	-0.61
7.23	8.86	-0.88
5.30	6.25	0.18
6.35	7.90	-0.16
7.15	8.96	-0.63

Intercept	Slope	
b_0	b_1	SSE
-4	1.9	3.4572

Figure 10.2.1. Visual Method
[Visual in Simple Regression]

10.3 The Solver Method

10.3.1 The Method

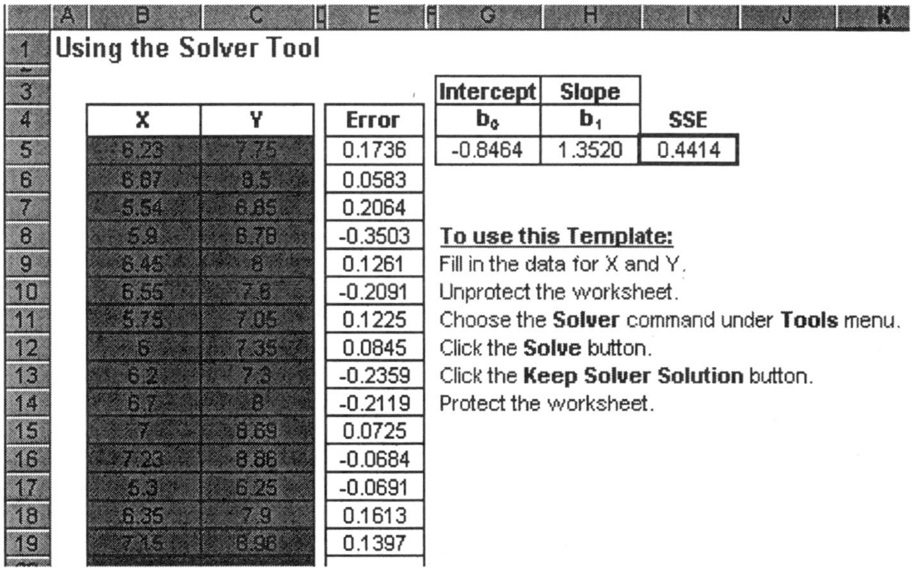

Using the Solver Tool

X	Y	Error
6.23	7.75	0.1736
6.87	8.5	0.0583
5.54	6.85	0.2064
5.9	6.78	-0.3503
6.45	8	0.1261
6.55	7.8	-0.2091
5.75	7.05	0.1225
6	7.35	0.0845
6.2	7.3	-0.2359
6.7	8	-0.2119
7	8.69	0.0725
7.23	8.86	-0.0684
5.3	6.25	-0.0691
6.35	7.9	0.1613
7.15	8.96	0.1397

Intercept	Slope	
b_0	b_1	SSE
-0.8464	1.3520	0.4414

Figure 10.3.1 The Solver Method

[Solver in Simple Regression]

In this method, we invoke the Solver tool to minimize SSE. Figure 10.3.1 shows the sheet named "Solver" in Simple Regression. After the data is entered in the shaded region, use the Solver tool to minimize SSE. The Solver is already set on the template to minimize SSE by changing Slope and Intercept. In order to use the Solver, it is necessary to unprotect the sheet using the **Protection...** command under the **Tools** menu. Once the Solver solution is obtained, protect the sheet again for safety.

10.3.2 Advantages and Disadvantages

The advantages of this method are:

1. Additional constraints can be imposed and still optimal solutions can be found quickly. For instance, suppose we want the slope to be not more than, say, 5. We can impose this constraint on the Solver and get the optimal solution for slope and intercept.
2. The user is conscious of what the calculation is achieving.

The disadvantages of this method are:

1. When data is changed, the intercept and the slope will not update automatically. The Solver must be invoked again.
2. Additional quantities will have to be calculated for further analysis.

10.4 LINEST Method

10.4.1 The Method

LINEST stands for LINear ESTimation. It is a built-in function in Excel to conduct regression analysis, and it is quite efficient. The general format for this function is

`LINEST(Y-values, X-values, Logical 1, Logical 2).`

The first two arguments, *Y*-values and *X*-values, refer to the ranges of cells containing them. Note that the *Y*-values comes first and not *X*. The third argument shown as Logical 1, if TRUE, tells the computer not to force the intercept to equal zero. If it is set as FALSE, the intercept will be forced to equal zero. The fourth argument shown as Logical 2, if TRUE, returns eight other statistics besides intercept and slope. If you have to use LINEST function but are not sure of its arguments, use the function wizard.

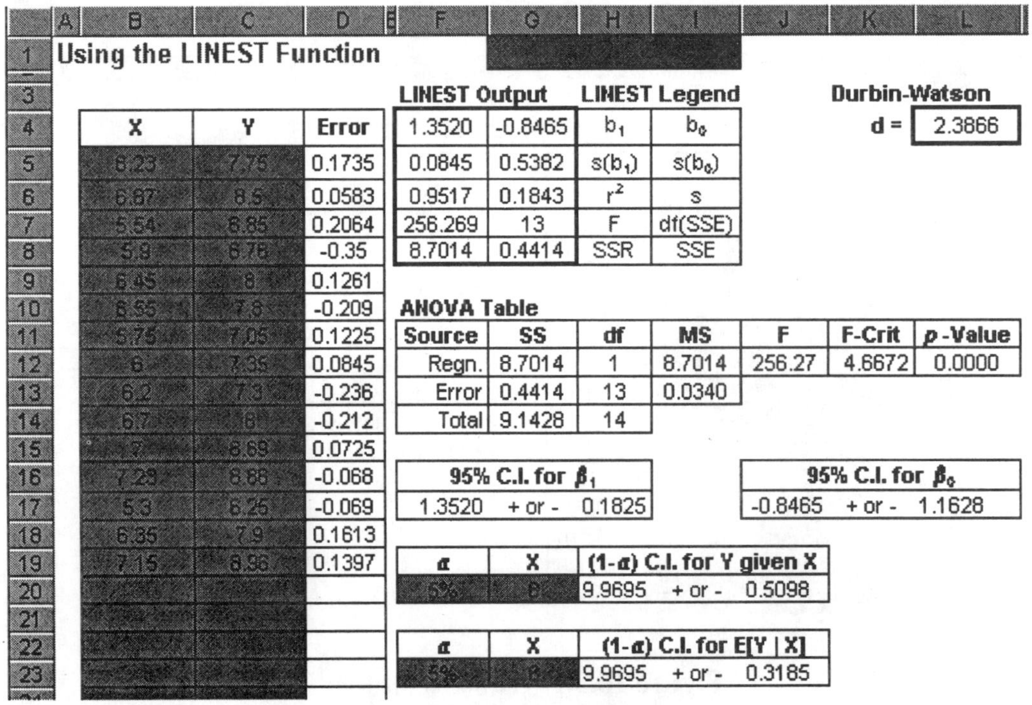

Figure 10.4.1 The LINEST Method
[LINEST in Simple Regression]

Figure 10.4.1 shows the sheet named "LINEST" where the LINEST function has been used to calculate the intercept and the slope. In the range F4:G8, we see the output from LINEST. Unfortunately, it does not output a legend. Hence a legend has been manually added in the range H4:I8.

10.4.2 Advantages and Disadvantages

The advantages of this method are:

1. It is fast, easy and accurate.
2. It provides additional statistics useful for further analysis.
3. If the data is changed, the results update automatically.
4. The additional constraint of intercept equals zero can be imposed by setting the third argument as FALSE.

The disadvantages are:

1. No other constraints except intercept equals zero can be imposed.
2. To a student just learning the technique, this method will be mostly a black box.

The advantages of this method are impressive. Hence we shall work with this method again to see further theory about regression.

10.5 The Matrix Method

10.5.1 The Method

The Matrix Method

MatX		MatY	Error
1	6.23	7.75	0.174
1	6.87	8.5	0.058
1	5.54	6.85	0.206
1	5.9	6.78	-0.35
1	6.45	8	0.126
1	6.55	7.8	-0.209
1	5.75	7.05	0.123
1	6	7.35	0.084
1	6.2	7.3	-0.236
1	6.7	8	-0.212
1	7	8.69	0.072
1	7.23	8.86	-0.068
1	5.3	6.25	-0.069
1	6.35	7.9	0.161
1	7.15	8.96	0.14

b	-0.8465	b_0	*s*(b)	0.5382	$s(b_0)$	
	1.35201	b_1		0.0845	$s(b_1)$	t 16.008

ANOVA Table

Source	SS	df	MS	F	F-Crit	*p*-Value
Regn.	8.70136	1	8.70136	256.27	4.6672	0.0000
Error	0.4414	13	0.03395			
Total	9.14276	14				

Durbin-Watson d = 2.3866

α	(1-α) CI for β_0		
5%	-0.8465	+ or -	1.16278

α	(1-α) CI for β_1		
5%	1.35201	+ or -	0.18246

Given x: 1, 8

α	(1-α) CI for y for given x		
5%	9.96952	+ or -	0.50979

α	(1-α) CI for E[y] for given x		
5%	9.96952	+ or -	0.31846

Figure 10.5.1 The Matrix Method
[Matrix in Simple Regression]

Figure 10.5.1 shows the sheet named "Matrix" where the matrix method has been employed. Here MatX, denoted by **X**, is obtained by adding a column of 1's in front. Thus it has size *n* x 2 where *n* is the sample size. MatY, denoted by **Y**, is simply the Y-values and has size *n* x 1. The errors form a matrix of size *n* x 1, denoted by ε. Let **b** denote the matrix $\begin{bmatrix} b_0 \\ b_1 \end{bmatrix}$. With these notations, we have

$$\mathbf{Y} = \mathbf{X}\mathbf{b} + \varepsilon$$

$$SSE = \varepsilon'\varepsilon = (\mathbf{Y} - \mathbf{Xb})'(\mathbf{Y} - \mathbf{Xb}) = \mathbf{Y'Y} - 2\mathbf{b'X'Y} + \mathbf{b'X'Xb}$$

The objective is to minimize SSE by changing **b**, and therefore we should have the first derivative of SSE with respect to **b** equal to zero. Doing that we get,

$$\mathbf{b} = (\mathbf{X'X})^{-1}\mathbf{X'Y}.$$

The above equation is all that one needs to find intercept and slope. In Excel, TRANSPOSE transposes a matrix, MMULT multiplies two matrices and MINVERSE calculates the inverse of a matrix. These three functions are sufficient to implement the above formula.

Let us assume that the data has been entered and the matrices have been named MatX and MatY. To calculate **b**,

- Select a 2 x 1 range of cells (such as H3:H4 in Figure 10.5.1.)
- Enter the formula `=MMULT(MINVERSE(MMULT(TRANSPOSE(MatX),MatX)),MMULT(TRANSPOSE(MatX),MatY)).`
- Press `CTRL+SHFT+Enter`

The intercept and slope appear in the selected range. Similar calculations have been employed in the template. To use the template, one just enters the *X* and *Y* data in the shaded area.

10.5.2 Advantages and Disadvantages

The <u>advantages</u> of this method are:

1. Matrix formulas are very compact and therefore easy to implement.
2. This method is the best way to calculate some advanced statistics such as the covariance of b_0 and b_1.
3. When the data is changed the results update automatically.
4. To impose the constraint intercept equals zero, all that needs to be done is eliminate the first column of 1's in MatX.

The <u>disadvantages</u> are:

1. Though not impossible, it is very difficult to impose constraints other than intercept equals zero.
2. The user needs to know matrix algebra.

10.6 The Analysis ToolPak Method

10.6.1 The Method

The sheet named "Analysis ToolPak" shows the implementation of this method. To implement the method, select the **Data Analysis ...** command under the **Tools** menu. In the dialog box that appears select **Regression**. A regression dialog box appears. The only new aspects are the entries in the **Residuals** section. Residuals are nothing but errors. In almost any regression analysis, the errors are important as well, and we will see more about that in the section on further theory. Selecting **Residuals** box causes errors to be calculated and displayed. Selecting the **Standardized Residuals** box causes the *z*-scores of errors to be output.

In the **Input** section, checking the box **Constant is Zero** imposes the condition intercept equals zero. The **Confidence Level** box is meant for computing confidence intervals of the slope and the intercept.

In the **Normal Probability** section of the dialog box, if one checks the **Normal Probability Plots** box one gets a plot of the *Y*-values vs. percentile ranks. This is really not a normal probability plot, and therefore not as useful.

10.6.2 Advantages and Disadvantages

The advantages of this method are:

1. In one stroke the user is able to get almost all the required results, including residual plots.
2. The constraint of intercept equals zero can be imposed.

The disadvantages are:

1. If the data is changed, the results will not update. It is necessary to invoke the macro again.
2. No other constraints besides intercept equals zero can be imposed.
3. To a student learning regression for the first time, this method is a black box.

10.7 The Normal Equations Method

10.7.1 The Method

Figure 10.7.1 shows the sheet named "Normal Eqns.". When regression problems are solved manually this is the method followed. We have SSE = $\Sigma e^2 = \Sigma (y - b_0 - b_1x)^2$. Since we are minimizing SSE by changing b_0 and b_1, we must have $\frac{\partial SSE}{\partial b_0} = 0$ and $\frac{\partial SSE}{\partial b_1} = 0$. The first condition $\frac{\partial SSE}{\partial b_0} = 0$ yields our first *normal equation* **NE1**: $\Sigma y = nb_0 + b_1 \Sigma x$

Similarly, $\frac{\partial SSE}{\partial b_1} = 0$ yields the second normal equation **NE2**: $\Sigma xy = b_0 \Sigma x + b_1 \Sigma x^2$

Since **NE1** and **NE2** are two simultaneous linear equations with two unknowns b_0 and b_1, they can be solved for b_0 and b_1. This solution of simultaneous equations is further simplified by the quantities SS_X, SS_Y and SS_{XY} defined as

$$SS_X = \Sigma x^2 - (\Sigma x)^2/n$$
$$SS_Y = \Sigma y^2 - (\Sigma y)^2/n$$
$$SSXY = \Sigma xy - (\Sigma x)(\Sigma y)/n$$

With these definitions,

$$b_1 = SS_{XY}/SS_X$$
$$b_0 = \bar{y} - b_1\bar{x}$$

where $\bar{x}$ and $\bar{y}$ are the respective sample means. These are the formulas implemented in the template.

Normal Equations Method

	X	Y	X^2	Y^2	XY
Sum	95.22	116.04	609.22	906.83	743.06
	6.23	7.75	38.8129	60.0625	48.2825
	6.87	8.5	47.1969	72.25	58.395
	5.54	6.85	30.6916	46.9225	37.949
	5.9	6.78	34.81	45.9684	40.002
	6.45	8	41.6025	64	51.6
	6.55	7.8	42.9025	60.84	51.09
	5.75	7.05	33.0625	49.7025	40.5375
	6	7.35	36	54.0225	44.1
	6.2	7.3	38.44	53.29	45.26
	6.7	8	44.89	64	53.6
	7	8.69	49	75.5161	60.83
	7.23	8.86	52.2729	78.4996	64.0578
	5.3	6.25	28.09	39.0625	33.125
	6.35	7.9	40.3225	62.41	50.165
	7.15	8.96	51.1225	80.2816	64.064

n			
15			
SS_X	SS_Y	SS_{XY}	
4.7602	9.1428	6.4359	
b_0	b_1		r
-0.8465	1.352		0.9756

ANOVA Table

Source	SS	df	MS
Regn.	8.7014	1	8.7014
Error	0.4414	13	0.034
Total	9.1428	14	

95% C.I. for b1		
1.352	+ or -	0.1825

Figure 10.7.1. Normal Equations Method
[Normal Eqns. in Simple Regression]

10.7.2 Advantages and Disadvantages

The <u>advantage</u> of this method is:

1. Students learning the technique for the first time get to see the details of the calculations better.

The <u>disadvantages</u> are:

1. All further statistics need further elaborate calculations.
2. Very difficult to impose constraints, except intercept equals zero. [When intercept is constrained to be zero, the formula $\Sigma xy/\Sigma x^2$ gives the optimal slope.]
3. The method is tedious and slow.

10.8 The Trendline Method

This method is good if one needs only a graph showing the regression line, and a few other simple statistics. Figure 10.8.1 shows the sheet named "Trendline". To create the trendline after drawing a scatter plot (as described in the introductory section),

- Click on any one of the data points in the graph. All the data points are highlighted.
- Click on the **Trendline...** command under the **Insert** menu.
- In the dialog box that appears, select the **Options** tab.
- At the bottom, check the **Display Equation on Chart** box
- Check the **Display R-squared Value on Chart** box
- Click the **OK** button.

A trendline appears on the graph. If you want to format this line further, select the **Selected Trendline...** command under the **Format** menu. You can now change the color and thickness of the line. You may also re-position the equation text anywhere on the plot.

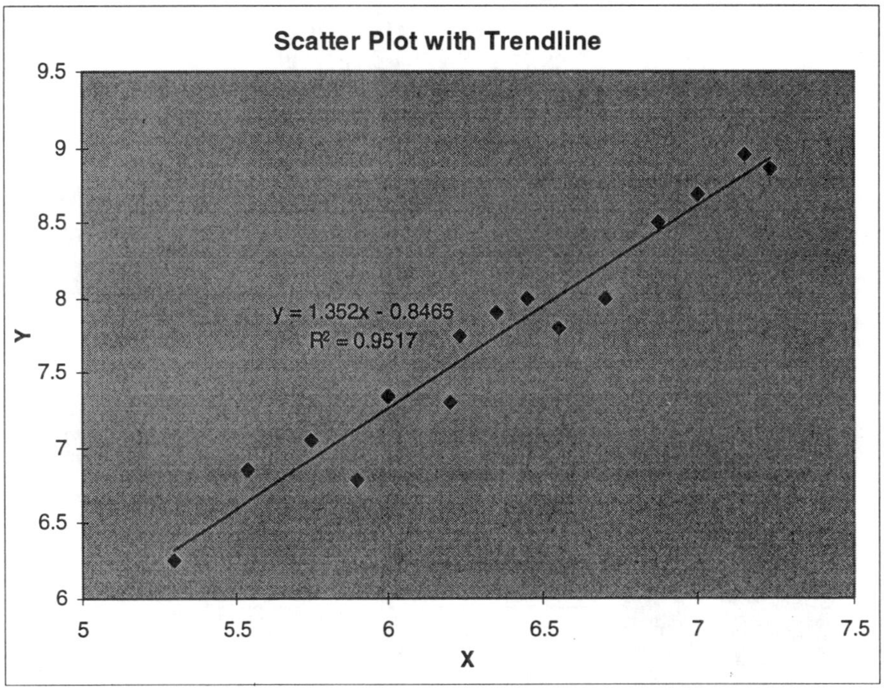

Figure 10.8.1 Scatter Plot with Trendline
[Trendline in Simple Regression]

The advantages of this method are:

1. It produces a nice graph with the regression line in it.
2. When data is changed the graph updates itself immediately.
3. It can calculate and display the regression equation and r^2 value right on the graph.

The disadvantages of the method are:

1. To a student learning the regression for the first time, this method is a black box.
2. The only constraint it can impose is intercept equals zero.
3. It does not provide any further statistics.

10.9 Regression Formulas

Many of the regression results can be calculated using simple built-in functions in Excel. The use of these functions results in compact one-line formulas for all the popular regression statistics. Figure 10.9.1 shows the sheet named "Formulas". This template shows the formulas used.

Regression Formulas

Row	XData	YData		F	G
3			n	15	=COUNT(XData)
4	6.23	7.75	Intercept	-0.8465	=INTERCEPT(YData,XData)
5	6.87	8.50	Slope	1.35201	=SLOPE(YData,XData)
6	5.54	6.85	Slope	1.35201	=LINEST(YData,XData)
7	5.90	6.78	$b_1 \mid (b_0 = 0)$	1.21969	=LINEST(YData,XData,FALSE)
8	6.45	8.00	r^2	0.9517	=RSQ(YData,XData)
9	6.55	7.80	r	0.9756	=CORREL(XData,YData)
10	5.75	7.05	r	0.9756	=PEARSON(XData,YData)
11	6.00	7.35	SSR	8.70136	=DEVSQ(TREND(YData,XData))
12	6.20	7.30	SSE	0.4414	{=SUMSQ(YData-TREND(YData,XData))}
13	6.70	8.00	SST	9.14276	=DEVSQ(YData)
14	7.00	8.69	SS_X	4.76024	=DEVSQ(XData)
15	7.23	8.86	SS_Y	9.14276	=DEVSQ(YData)
16	5.30	6.25			
17	6.35	7.90	When x =	8	
18	7.15	8.96	y-hat =	9.96952	=TREND(YData,XData,F13)

Figure 10.9.1. One-Line Formulas for Regression Statistics
[Formulas in Simple Regression]

Some points to note in this figure are:

1. The use of INTERCEPT and SLOPE formulas in cells F4 and F5.
2. LINEST used in a single cell, returns just the slope.
3. If the third argument is False, LINEST returns the slope when intercept equals zero.
4. CORREL and PEARSON return the same result, namely, the correlation coefficient.
5. In the absence of the third argument, TREND returns the $\hat{y}$ value corresponding to each x-value in the data. If a third argument is present, it is taken as the given x, and a corresponding $\hat{y}$ value is returned (as in cell F18). If the fourth argument is included and specified as FALSE, the intercept is constrained to equal zero.
6. SUMSQ returns the sum of the squares of all the arguments.

The advantages of this method are:

1. Compact formulas return most of the popular regression statistics. Therefore, this method is appropriate if all we need are a few statistics, and not all the details.
2. When data is changed, the results update automatically.

The disadvantages are:

1. To a beginning student, this method is a black box.
2. No constraint, other than intercept equals zero, can be imposed.
3. Advanced statistics will need further calculations.

10.10 Recommended Method

Since there are as many as seven different methods, it is useful to know which ones are suitable under which conditions. Considering the advantages and disadvantages of all these methods, the author recommends that a beginning student start with the Visual method and then move on to the Normal Equations method and then to the LINEST method. If the user knows matrix algebra, then he/she may skip the LINEST method and go on to the Matrix method. Use the Trendline method only if the graph is all that is needed.

10.11 Further Theory

10.11.1 Correlation

Often we need to know the expected value of the product of two random variables X and Y. A little algebra proves

$$E[XY] = E[X]E[Y] + E[(X - \mu_x)(Y - \mu_y)].$$

The second term $E[(X - \mu_x)(Y - \mu_y)]$ is defined as the *covariance* of X and Y, denoted by Cov(X, Y). Thus we get

$$E[XY] = E[X]E[Y] + \text{Cov}(X, Y).$$

The covariance tells us how X and Y are related. If Y is high whenever X is high, then they would have a positive covariance; if Y is high whenever X is low, they would have a negative covariance. Advertising and sales usually have positive covariance and price and sales usually have negative correlation.

Covariance is also needed for calculating the variance of the sum of two random variables, as seen in the following formula:

$$V(X + Y) = V(X) + V(Y) + 2\,\text{Cov}(X, Y).$$

If X is measured in dollars and Y in meters, the Cov (X, Y) would be in dollar-meters, and this unit may be cumbersome. To turn it into a dimensionless number it is divided by $\sigma_x\,\sigma_y$ and the result is called the *coefficient of correlation* between X and Y. The coefficient of correlation is denoted by ρ_{xy} or simply ρ. Thus we have,

$$\rho = \frac{Cov(X,Y)}{\sigma_x \sigma_y}.$$

By a famous theorem called *Cauchy-Schwartz inequality*, the absolute value of the numerator Cov (X, Y) cannot exceed the denominator $\sigma_x\,\sigma_y$ and therefore, the absolute value of ρ cannot exceed 1. A ρ value of 1 denotes perfect positive correlation and a value of -1 denotes perfect negative correlation. When we conduct a regression analysis using sample data, the quantity

$$r = \frac{SS_{XY}}{\sqrt{SS_X\, SS_Y}}$$

is an unbiased estimate of ρ. If we use LINEST function for the regression, one of the results that LINEST provides is r^2. See cell F7 in Figure 10.4.1. The square root of this number yields r.

Why does LINEST provide r^2 rather than r? The reason is that r^2 has an important significance and simpler interpretation. It signifies the proportion of variance in the dependent variable Y that is (explained, caused or) *determined* by the independent variable X. Hence r^2 is called the *coefficient of determination*. If $r^2 = 1$, then all of the variance in Y is determined by X, and all the data points would fall exactly on the regression line. As r^2 decreases, there will be more and more errors and the data points stray farther and farther from the regression line.

10.11.2 ANOVA and the *F*-Test

The assumptions that enable t and F tests are repeated here for convenience.

A1. $Y = \beta_0 + \beta_1 X + \varepsilon$

where β_0 denotes the intercept, β_1 the slope and ε the error.

A2. The error ε is normally distributed with zero mean and finite variance σ^2. The variance σ^2 is constant for all values of X. Further, the errors at one value of X is not correlated with those at any other value of X.

A3. The error ε is not correlated with X.

A4. The X values of the sample are fixed and not random

Under these assumptions, it is possible to find what part of SS_Y was caused by X. The rest is caused by error, and it is nothing but the SSE. We organize this analysis in a standard ANOVA table. (E.g.: See the ANOVA table in Figure 10.4.1.) Under the Source column, the effect of X is labeled Regression and the value itself is found under the SS column. This is referred to as SSR which stands for Sum of Squares due to Regression. SSR can be calculated using

$$SSR = b_1^2 SS_X.$$

On the "Normal Eqns." sheet, SS_X has been directly calculated and used in the above equation. On other sheets, SS_X can be calculated using the formula `=DEVSQ(X)`. If LINEST function is used, SSR itself is output as one of the results.

The total SS, denoted by SST, is a measure of the variance in *Y*. It is the same as the SS_Y calculated in the "Normal Eqns." sheet. It is also the sum of SSR and SSE. Another way to calculate it is to use the formula `=DEVSQ(Y)`.

Coming to degrees of freedom, the Y values have *n* degrees of freedom. The degrees of freedom for SST is thus $(n - 1)$. The $\hat{Y}$ values predicted by the regression line have two degrees of freedom since we are free to choose two quantities, namely, the intercept and the slope. Because the errors are generated by subtracting the $\hat{Y}$ values from the *Y* values SSE has $(n - 2)$ degrees of freedom. See the diagram below.

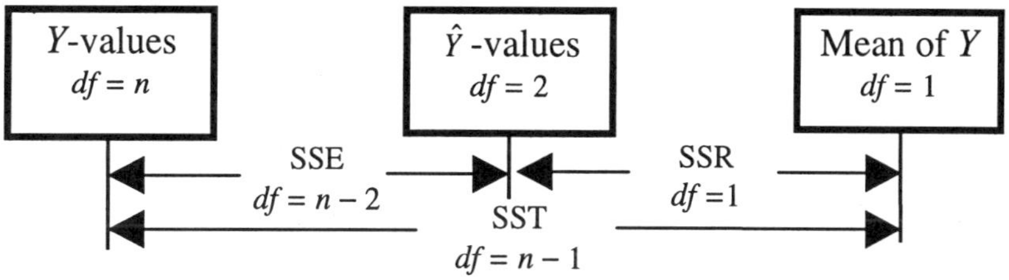

The rest of the ANOVA table is similar to what we saw in the previous chapter. Under the assumptions **A1-A4**, the ration MSR/MSE should follow an *F*-distribution. Hence we can conduct an *F*-test for the null hypothesis H_0: *X* does not affect *Y* versus H_1: *X* affects *Y*.

In the current example, the *F* ratio is 256 with a *p*-value of almost zero. Hence, we reject the null hypothesis. In other words, the *X* does affect *Y*.

MSE is the estimate of the error variance σ^2 and $\sqrt{\text{MSE}}$, denoted by *s*, is an unbiased estimate of σ. The LINEST function provides the value of *s* in its output.

10.11.3 *t*-Test

The most importance quantity calculated in a regression analysis is the slope, because it tells us exactly how much effect *X* has on *Y*. If the slope is 5, it means that every 1 unit increase in *X* causes 5 units increase in *Y*. If the slope is zero, the it means *X* does not affect *Y* at all. The above hypotheses can also be written as

$$H_0\text{: Slope} = 0$$
$$H_1\text{: Slope} \neq 0$$

A *t*-test can be applied here. Under the assumptions **A1-A4**, the statistic $b_1/s(b_1)$ where $s(b_1)$ is the standard deviation of b_1 follows a *t*-distribution. $s(b_1)$ is calculated using the formula

$$s(b_1) = \frac{s}{\sqrt{SS_X}}.$$

The LINEST function provides $s(b_1)$ as one of the results. In the matrix method the matrix

$$s^2(\mathbf{b}) = \begin{bmatrix} s^2(b_0) & Cov(b_0, b_1) \\ Cov(b_0, b_1) & s^2(b_1) \end{bmatrix}$$

can be calculated using the formula $s^2(\mathbf{b}) = \text{MSE}\,(\mathbf{X'X})^{-1}$. Thus the matrix method is quite powerful.

In the current example the t ratio is 16, and therefore the null hypothesis is rejected. This test is equivalent to the *F*-test in that if one of them rejects the null hypothesis the other one will reject it too. Thus there must be a relationship between the *t* ratio and the *F* ratio. The relationship is $F = t^2$. In the current example we have $F = 256 = 16^2 = t^2$.

The knowledge of $s(b_1)$ also helps us to calculate a $(1 - \alpha)$ confidence interval for β_1 given by the formula $b_1 \pm t_{\alpha/2}\, s(b_1)$. This has been done in the LINEST sheet in the range F18:H18 in Figure 10.4.1. Similar calculation can also be done for β_0, but it does not have as much significance.

10.11.4 Prediction

Another important use of regression is the prediction of Y when the X value is known. The predicted value of Y is denoted by $\hat{Y}$ and is calculated as

$$\hat{Y} = b_0 + b_1 X.$$

But this is only a point estimate and we would rather have a confidence interval. To compute a confidence interval, we need to know the variance of $\hat{Y}$ denoted by $V(\hat{Y})$. Before we see the formula for $V(\hat{Y})$, let us compare the formula for $\hat{Y}$ above with the regression model

$$Y = \beta_0 + \beta_1 X + \varepsilon .$$

The differences are:
1. The error term is missing.
2. Instead of β_0 we have b_0 which is an estimate of β_0, and is subject to sampling error.
3. Instead of β_1 we have b_1 which is an estimate of β_1, and is subject to sampling error.

The variance of $\hat{Y}$ is given by the formula

$$V(\hat{Y}) = \sigma^2 + \sigma^2/n + \sigma^2(X - \mu_x)^2/SS_X$$

where σ^2 is the error variance. This formula has three terms that correspond to the variance contributed by the three differences above. Now, we do not know μ_x and σ^2. In place of μ_x we substitute the sample mean $\bar{x}$. In place of σ^2 we substitute MSE, because under the assumptions **A1-A4**, MSE is an unbiased estimate of σ^2. The square root of MSE is denoted by s. Thus, a $(1 - \alpha)$ confidence interval for $\hat{Y}$ when $X = x$ is

$$\hat{Y} \pm t_{\alpha/2}\, s \sqrt{1 + \frac{1}{n} + \frac{(x - \bar{x})^2}{SS_X}} .$$

This formula has been implemented in the templates wherever confidence intervals for predictions are needed. With matrices, this expression is a little more compact. If we denote the matrix $\begin{bmatrix} 1 \\ x \end{bmatrix}$ by $\mathbf{x}$, then the confidence interval is given by

$$\hat{Y} \pm t_{\alpha/2}\, s \sqrt{1 + \mathbf{x}'(X'X)^{-1}\mathbf{x}}$$

This formula has been implemented in the sheet named "Matrix".

At times, we need the confidence interval for the expected value of $\hat{Y}$. In this case, the effect of the error term goes to zero, and therefore the 1 inside the radical sign drops out. We have the formula for the confidence interval for $E[\hat{Y}]$ as

$$\hat{Y} \pm t_{\alpha/2}\, s \sqrt{\frac{1}{n} + \frac{(x - \bar{x})^2}{SS_X}}$$

and with matrix notation,

$$\hat{Y} \pm t_{\alpha/2}\, s \sqrt{\mathbf{x}'(X'X)^{-1}\mathbf{x}} .$$

10.11.5 Residual Analysis

Assumption **A2** we made about the error in the regression model is important for the F and t tests. It is therefore advisable to check how good that assumption is at the end of a regression analysis when the errors have been calculated. We shall see three actions we may take in this regard. Assumption **A3** is also about the errors, but we need not test if it is satisfied because it is *automatically* satisfied as we minimize SSE.

The first is the visual analysis. Here we look at a scatter plot of the errors and see if the plot is random or if it shows any definite pattern that suggests a violation of the assumption. Figure 10.11.1 shows the scatter plot of the errors vs. Y, taken from the LINEST sheet.

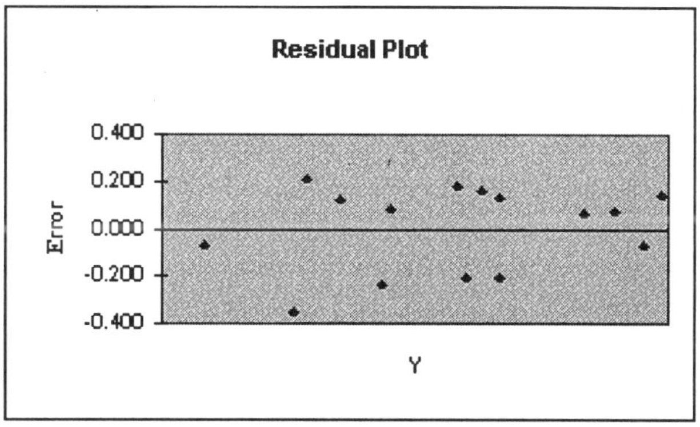

Figure 10.11.1 A Residual Plot

This plot does not show any pattern in the error and hence it may be declared random. If the error plot shows a pattern, then our assumption that the errors are random is violated and the hypothesis tests are invalid. Two non-random patterns are shown in Figure 10.11.2. The one on the left shown non-linearity in the relationship between X and Y. The one on the right shows *heteroscedasticity*, meaning the error variance is not constant.

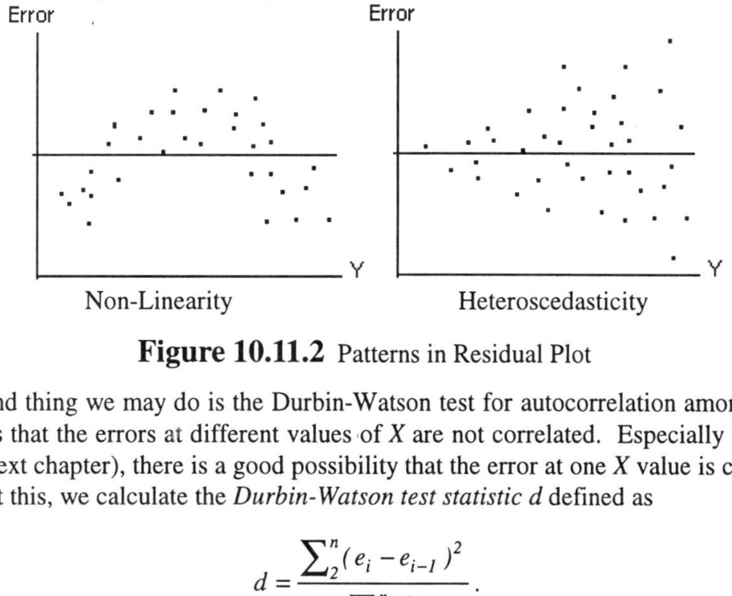

Figure 10.11.2 Patterns in Residual Plot

The second thing we may do is the Durbin-Watson test for autocorrelation among the errors. Our assumption (**A2**) is that the errors at different values of X are not correlated. Especially in time-series data (discussed in the next chapter), there is a good possibility that the error at one X value is correlated with that at the next. To test this, we calculate the *Durbin-Watson test statistic d* defined as

$$d = \frac{\sum_{2}^{n}(e_i - e_{i-1})^2}{\sum_{1}^{n} e_i^2}.$$

This statistic d varies from 0 to 4 with mean 2. If the calculated value is 2 or close to 2, it means there is no autocorrelation and our assumption is valid. As it goes toward zero or 4, the amount of autocorrelation increases and our assumption becomes questionable. A d value of 0 indicates perfect positive autocorrelation and 4 indicates a perfect negative autocorrelation. The exact values of d at which one would declare the assumption invalid depends on the sample size, and these values may be found in tables in textbooks. For the example problem solved on the LINEST template $d = 2.3866$ (see cell L5 in Figure 10.4.1). This indicates some negative autocorrelation but for a sample size of 15, this is not large enough to invalidate our assumption of no autocorrelation.

The third test we may conduct on residuals is the test for normality. There are several ways to test normality, and we shall see just one, namely, the normal probability plot. In this plot, we compare the errors with the standard normal distribution and see how they fit. To do this, we first calculate the quantile of each error value by calculating its rank and dividing it by $(n + 1)$. Then we compute the z-value corresponding to that quantile using the NORMSINV function. We then plot the z values against the error values and see if they fall roughly along a straight-line. The more the points deviate from a straight-line the more non-normality is there in the errors. For our example problem, a normal probability plot has been constructed in the sheet named "Norm. Prob. Plot". It is shown here in Figure 10.11.3. The points do deviate from the straight-line making the assumption of normality questionable.

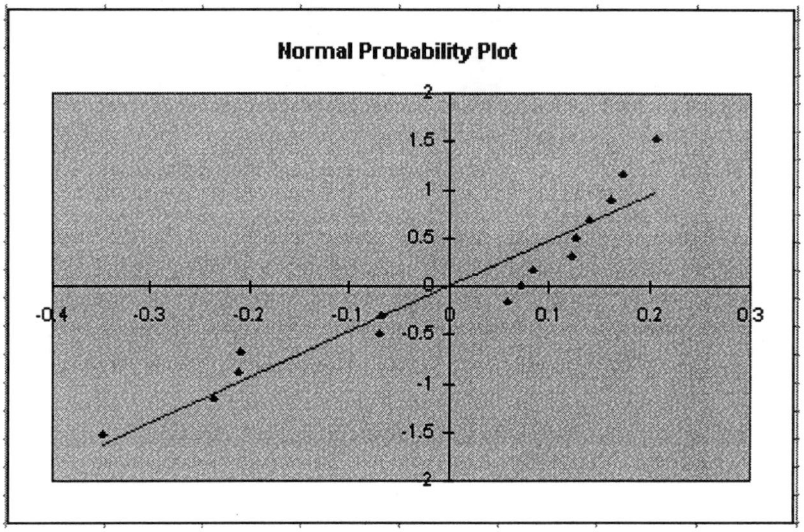

Figure 10.11.3 Normal Probability Plot
[Norm. Prob. Plot in Simple Regression]

10.12 Exercises

1. Consider the data in Exercise 10-13 in Aczel's book.
i. Do the series of exercises 10-13, 10-23, 10-28, 10-39, 10-48, 10-57, 10-67 from the book.
ii. Give 95% and 99% confidence interval for Y and E[Y] when $X = 50$.
iii. Conduct a Durbin-Watson test on the residuals.
iv. Construct a normal probability plot of the residuals.

2. Do Case 10 in Aczel's book.

3. According to the Capital Asset Pricing Model (CAPM), the risk associated with the returns from a capital asset is proportional to the slope β_1 (or simply β) obtained by regressing the asset's past (annual) returns with the corresponding returns of the average portfolio called the *market portfolio*. [The return of the *market portfolio* represents the return earned by the average-investor, which is a weighted average of the returns from all the assets in the market.] Larger the slope β of an asset, larger is the risk associated with that asset. A β of 1.00 represents average risk.

The returns from Philips Electronics shares traded in the Hong Kong stock market, and the corresponding returns for the Hong Kong market portfolio for the past 12 years are given below.

Market Return(%)	16.02	12.17	11.48	17.62	20.01	14.00	13.22	17.79	15.46	8.09	11.00	18.52
PhilipsReturn(%)	21.05	17.25	13.10	18.23	21.52	13.26	15.84	22.18	13.26	5.64	10.55	17.86

i. Carry out the regression and find the β for Philips. Give a 95% confidence interval for this β. Would you say Philips's risk is above average?
ii. If the market portfolio return for 1994 is 10%, what is the return on Philips predicted by the regression equation? Give a 95% confidence interval for this prediction.
iii. Conduct a t-test to test the null hypothesis: $\beta = 0$.
iv. Construct the ANOVA table for the regression.
v. Construct a residual plot. Conduct the Durbin-Watson test and normality test of the residuals.
vi. The "risk-free rate of return" is the rate associated with an investment which has no risk at all, such as lending money to the US government. For the Hong Kong market this rate is 6.2%. According to CAPM,

when the return from the market portfolio is equal to the risk-free rate, the return from every asset must also be equal to the risk-free rate. In other words, if the market portfolio return is 6.2%, then Philips's return should also be 6.2%. It implies that the regression line must pass through the point (6.2, 6.2). Repeat the regression forcing this constraint and recalculate β for Philips.
[*Hint*: *The formula* `=LINEST(Ydata-b,Xdata-a,FALSE,TRUE)` *will force the regression line to pass through the point with coordinates (a, b).*]

10.13 Projects

1. When Y is regressed against X, the slope of the regression line is 5. When X is regressed against Y, one might then expect the slope to be the reciprocal, 1/5. But this is rarely the case. Can you explain why?

2. Carry out the regression for the following data.

X	1	2	3	4	5
Y	13	18	23	28	33

i. What would happen to the intercept if we add 3 to all the Y values ? What if we add 4 ? Generalize your observation.
ii. What would happen to the slope if we add 3 to all the Y values ? What if we add 4 ? Generalize your observation.
iii. What would happen to the intercept if we multiplied all the Y values by 2 ? by 3 ? Generalize your observation.
iv. What would happen to the slope if we multiplied all the Y values by 2 ? by 3 ? Generalize your observation..
v. What would happen to the intercept if we did both multiplying by 2 and adding 3 to all the Y values ? Generalize your observation.
vi. What would happen to the slope if we did both multiplying by 2 and adding 3 to all the Y values ? Generalize your observation.
vii. Repeat questions i. through vi. for the same operations on X values.
viii. You may go further along these lines by changing both X and Y values at the same time. Try to answer this question: What would happen to the intercept and the slope if X values are changed to $aX + b$ and Y values are changed to $cY + d$ at the same time, where a, b, c, and d are some constants ?

11. Multiple Regression

11.1 Introduction

Often the dependent variable Y depends on more than one independent variable. It would be desirable to relate Y to all these variables in one equation. The regression analysis that relates a dependent variable to more than one independent variable is called *Multiple Regression*. The objective is to get an equation in the format

$$Y = \text{Intercept} + \text{Slope}_1 * X_1 + \text{Slope}_2 * X_2 + \ldots + \text{Slope}_k * X_k$$

where k is the number of independent variables X_1, X_2, … X_k that affect Y. Most of the concepts in multiple regression are the same as those of simple regression. For instance, the equation that best fits the data is taken to be the one that minimizes SSE. The important differences are that multiple regression is more calculation intensive and that the phenomenon of multicollinearity can complicate matters.

The assumptions behind the technique are:

A1. $Y = \beta_0 + \beta_1 X_1 + \beta_2 X_2 + \ldots + \beta_k X_k + \varepsilon$

A2. The error ε is normally distributed with zero mean and a constant variance σ^2. Errors at different values of the independent variables are independent of each other.

A3. The error ε is uncorrelated with every one of the independent variables.

A4. The sampled values of the independent variables are fixed, without errors.

The Excel functions LINEST and TREND work for both simple and multiple regression. The functions INTERCEPT, SLOPE, and RSQ work only for simple regression.

11.2 LINEST Method

We shall quickly see how LINEST function may be used for multiple regression. Figure 11.2.1 shows the sheet named "LINEST" in Multiple Regression workbook.

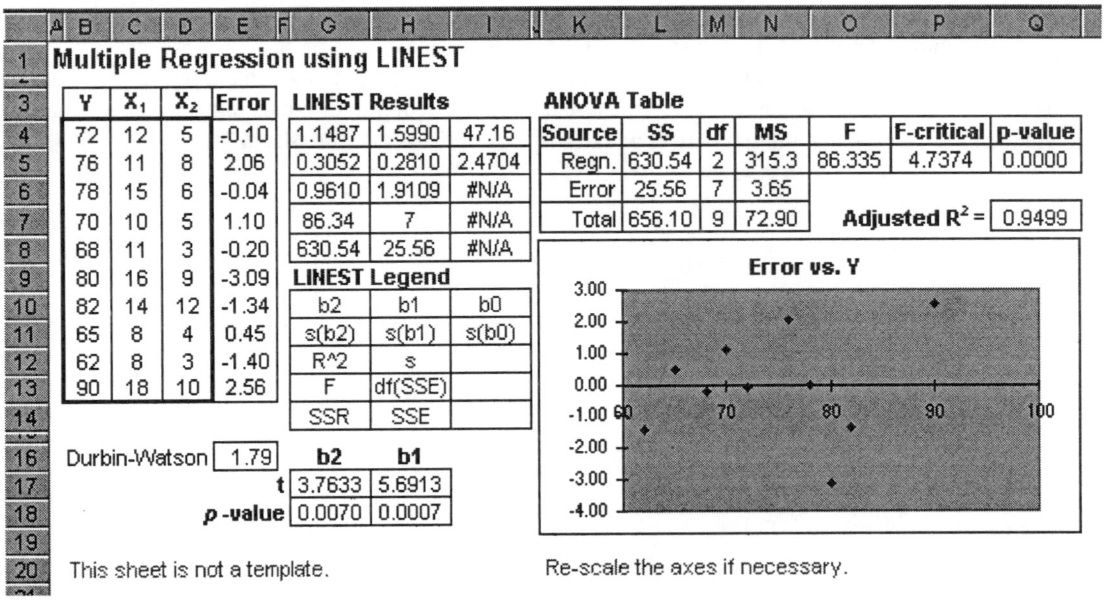
Multiple Regression using LINEST

Y	X_1	X_2	Error
72	12	5	-0.10
76	11	8	2.06
78	15	6	-0.04
70	10	5	1.10
68	11	3	-0.20
80	16	9	-3.09
82	14	12	-1.34
65	8	4	0.45
62	8	3	-1.40
90	18	10	2.56

LINEST Results

1.1487	1.5990	47.16
0.3052	0.2810	2.4704
0.9610	1.9109	#N/A
86.34	7	#N/A
630.54	25.56	#N/A

LINEST Legend

b2	b1	b0
s(b2)	s(b1)	s(b0)
R^2	s	
F	df(SSE)	
SSR	SSE	

ANOVA Table

Source	SS	df	MS	F	F-critical	p-value
Regn.	630.54	2	315.3	86.335	4.7374	0.0000
Error	25.56	7	3.65			
Total	656.10	9	72.90			

Adjusted R^2 = 0.9499

Durbin-Watson 1.79

	b2	b1
t	3.7633	5.6913
p-value	0.0070	0.0007

This sheet is not a template.

Re-scale the axes if necessary.

Figure 11.2.1 LINEST Method
[LINEST in Multiple Regression]

After entering the data in the range B4:D13 the following actions are taken to use the LINEST function:

- Select the range G4:I8. [This is a 5 x 3 area because the number of columns must be equal to the number of independent variables plus 1 for the intercept. The number of rows is always 5]
- Enter the formula `=LINEST(B4:B13,C4:D13,TRUE,TRUE)`
- Press `CTRL+SHFT+Enter`

LINEST provides the same statistics that we saw in the case of simple regression. The difference is that in the first two rows we get a slope and the standard error of the slope for each independent variable and also for the intercept. Thus there are more entries in the first two rows. The Legend gives the details of which output is which. Note the (reverse) order in which the slopes are displayed. The unused cells in the last three rows display #N/A which stands for Not Applicable.

The *t* ratio of each slope and the corresponding *p*-values appear in the range G17:H18. These ratios test the null hypothesis

H_0: The slope is zero.

In the ANOVA table, the *df* for SS due to Regression (SSR) is equal to the number of independent variables, or *k*, which is 2. The *df* for SS Total (SST) is $n - 1 = 10 - 1 = 9$. See the diagram below for degrees of freedom. In the diagram, the *df* for $\hat{Y}$ values is $(k + 1)$ because, including the intercept and *k* number of slopes, $(k + 1)$ values are chosen by us. The *F*-ratio tests the null hypothesis

H_0: All the slopes are zero.

If the *p*-value is small enough, as in the current example, the null hypothesis is rejected.

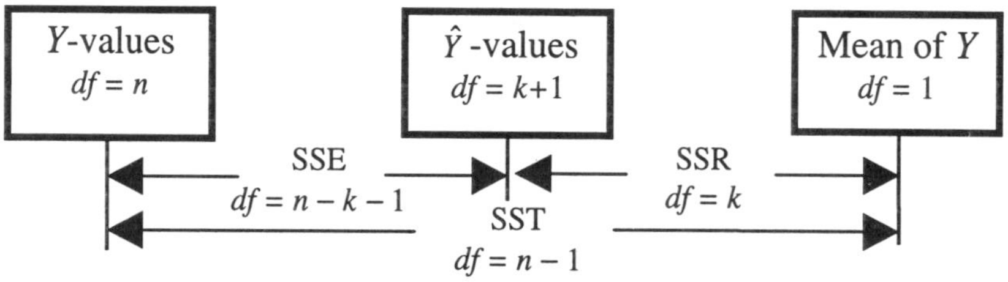

In simple regression the square root of the coefficient of determination r^2 was the correlation coefficient between *X* and *Y*. In the case of multiple regression the symbol for coefficient of determination is R^2 and its square root has no significance. Recall that $R^2 = \text{SSR/SST} = 1 - \text{SSE/SST}$

Suppose we have a regression with three independent variables and we add a fourth one. With the fourth slope constrained at zero we can achieve the same minimum SSE as before. With the additional freedom of changing the fourth slope we can, if anything, reduce the SSE further. Thus an addition of an independent variable into the model cannot decrease R^2. Even a nonsensical independent variable would increase R^2, albeit slightly. We therefore want to be wary of artificially increasing the R^2 by indiscriminately adding new independent variables. We do this by adjusting the R2 to account for the increased number of independent variables. The *adjusted* R^2 is defined as

$$\text{Adjusted } R^2 = 1 - \text{MSE/MST}$$

where MST = SST/*df*(SST). If the ANOVA table is completed MSE and MST values would be available from it, and adjusted R^2 can be calculated. Adjusted R^2 will always be less than R^2 and unlike R^2, it may decrease when a new dependent variable, especially one that is not well related to *Y*, is added to the model.

The errors have been calculated and plotted on the spreadsheet. The Durbin-Watson statistic *d* has been calculated in cell E16 using the formula

```
=SUMSQ(E5:E13-E4:E12)/SUMSQ(E4:E13).
```

11.3 The Solver Method

The Solver method has been carried out on the sheet named "Solver." The only difference here compared to simple regression is that there are more than one slope to calculate. A strong advantage of this method is that constraints, such as $b_1 + b_2 \leq 1$, can be imposed.

11.4 The Analysis ToolPak Method

This method works exactly the same way as simple regression. Hence it is not elaborated here.

11.5 The Matrix Template

The matrix template (comprising three sheets named "Data", "Results" and "Residuals" in Multiple Regression seen in Figures 11.5.1, 2 and 3) is the most convenient method for multiple regression. It can accommodate up to ten independent variables. The author recommends that a beginner use the Solver and LINEST methods to learn the concepts and then move to this method.

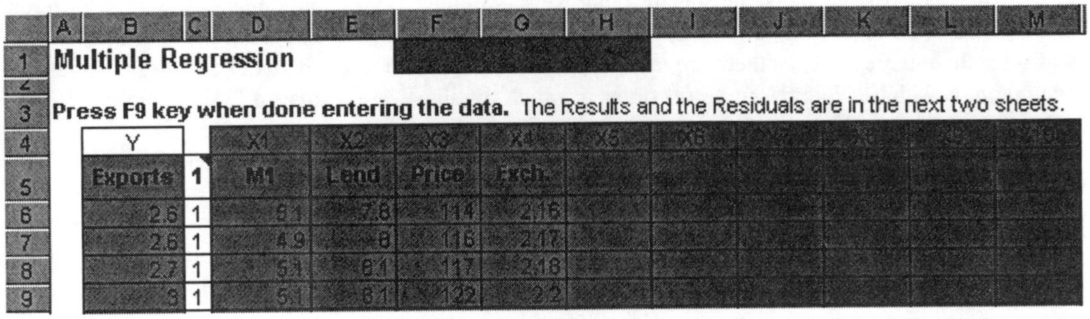

Figure 11.5.1. The Data Sheet
[Data in Multiple Regression]

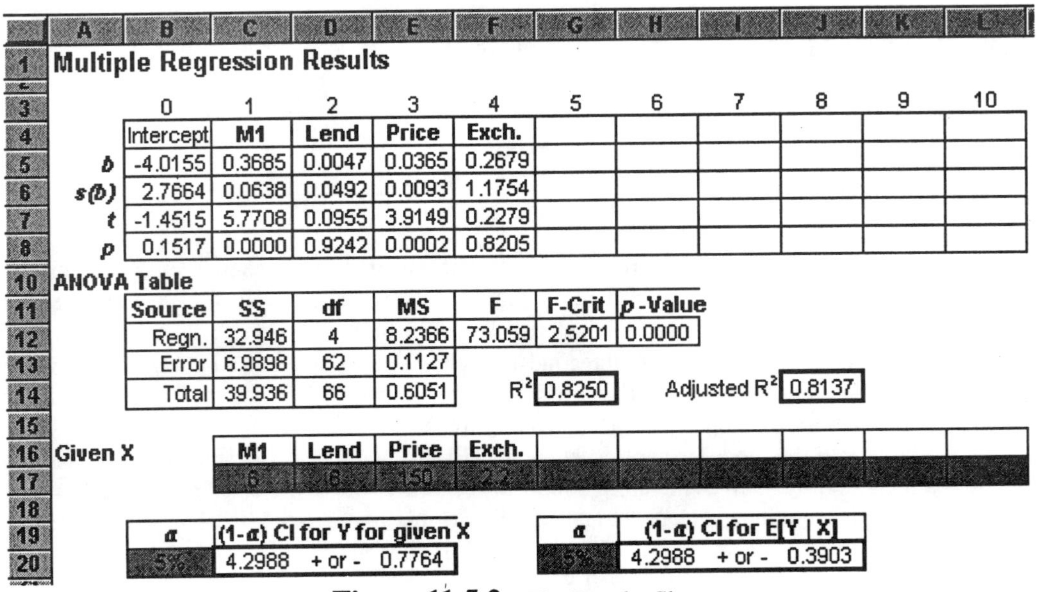

Figure 11.5.2. The Results Sheet
[Results in Multiple Regression]

To use the template,

- Open the sheet named **Data**.
- Clear any data already entered.
- Enter the new data for *Y* and all the *X* variables.
- Press the F9 key to start the calculation.
- Open the next sheet named **Results** to see the results.
- Open the next sheet named **Residuals** to see a residual analysis.
- If you want a normal probability plot, copy and transfer the Errors to the sheet named "Norm. Prob. Plot" in Simple Regression workbook.

The greatest advantage of this method is that the results include confidence intervals for predicted *Y*. When *all* the necessary *X* values are filled in on the Results sheet in the range C17:L17, the confidence intervals for *Y* and E[*Y*] appear in their respective places.

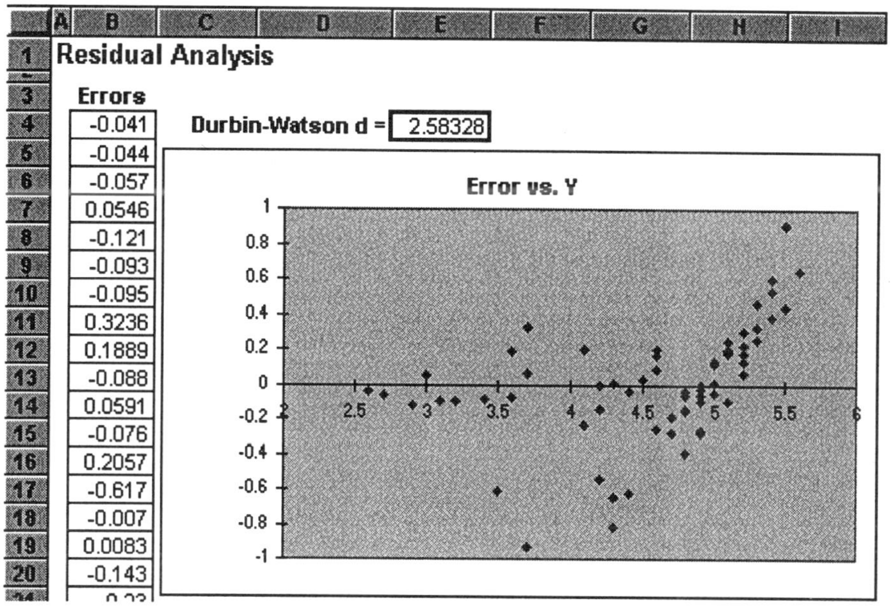

Figure 11.5.3 Residuals Sheet
[Residuals in Multiple Regression]

11.6 Multicollinearity

Because adding a new independent variable can increase R^2 and make the regression look better, there is a strong tendency among the users to keep adding new ones. One of the dangers in doing it is Multicollinearity among the independent variables.

Let us say we have conducted a multiple regression with two independent variables X_1 and X_2 and the regression equation is $Y = 1 + 5X_1 + 10X_2$. Suppose, as an experiment, we add a third independent variable X_3 which is identical to X_2, meaning the sample values of X_3 are identical to those of X_2. If we tried the regression using a spreadsheet template or another software, with X_3 included, we would get an error message such as "Warning: Division by zero" or "There is multicollinearity among the independent variables." Why? A little thought reveals that many equations, such as

$$Y = 1 + 5X_1 + 9X_2 + X_3$$
$$Y = 1 + 5X_1 + 8X_2 + 2X_3$$
$$Y = 1 + 5X_1 - 200X_2 + 210X_3$$

will all fit the data equally well. The calculation process stalls, and it is also difficult to interpret the regression equation. Is the slope for X_2 9, or 8, or –200, or what?

If X_3 was not identical to X_2, but almost identical to X_2, then the computation might not stall, and we would get some output. Say, we get the regression equation as $Y = 1 + 5X_1 + 8X_2 + 2X_3$. Now if we make a slight change in the data (anywhere in X_1, X_2, X_3 or Y), the regression equation might jump to the very different $Y = 1 + 5X_1 - 200X_2 + 210X_3$ signifying a serious problem.

The problem would exist even if X_3 was not related to X_1 or X_2 individually but together they were related, or almost related, as, say, $X_3 = 7.104X_1 - 3.896X_2$. When we have many more independent variables, the possibilities are numerous.

The presence of multicollinearity may not be obvious. Its symptoms are:

1. The slope of an independent variable is very different from what was expected. It may even happen that a negative value appears where a positive value is expected.
2. If a slight change is made in the data the slopes change a lot.
3. The standard errors of the slopes (reported by LINEST function for example) are too large.
4. The F-ratio may be significant proving that at least one slope is not zero. But the t-ratio for every one of the individual slopes may not be significant.

When multicollinearity is present, the most common way to alleviate the problem is to remove the offending independent variable(s) from the model. One way to do this is to regress each one of them with the rest of the independent variables and look at the R^2 value of the regression. If an R^2 value is 0.9 or more indicates possible problems of multicollinearity. Among those that are more than 0.9 the independent variable with the highest R^2 value should probably be thrown out of the model. This is tedious, but with the help of computers, it is probably not that bad.

11.7 Selection of Independent variables

In the interest of parsimony and to avoid multicollinearity, one would be careful in the selection of independent variables to be included in the model. This usually requires some trial and error. Suppose one has ten independent variables in the template and wants to choose the best four. For the template to work, the included variables should be on columns C, D, E and F. In other words, they should start at column C and be contiguous. To bring the four variables to the desired columns one may sort the columns of data as follows.

- Label the desired four variables as `X1`, `X2`, `X3` and `X4` on row 4 at the top
- Label the rest of the columns `X5` through `X10` in any order.
- Unprotect the sheet.
- Select the data in all the columns by dragging the mouse over the whole range.
- Select the **Sort...** command under the **Data** menu.
- In the dialog box, click the **Options** button.
- In the **Orientation** area, click the **Left to Right** button if it is not already clicked, and then click the **OK** button.
- On returning to the **Sort** dialog box, make sure the **Sort by** box says **Row 4**. If not, select it from the drop down list. Make sure the **Ascending** button is checked. Click the **OK** button.
- Protect the sheet.

The general principles of selecting the independent variables are:

1. Try to minimize the number of independent variables in the model without sacrificing too much in R^2.
2. Avoid multicollinearity. Make sure the slopes and their signs agree with intuition.
3. It is preferable that the *t*-ratios of all the slopes are significant.
4. That an included variable would indeed affect the dependent variable should appeal to common sense.

11.8 Exercises

1. Do Exercise 11-114 from Aczel's book.
2. Do Case 11 in Aczel's book.

12. Time Series, Forecasting and Index Numbers

12.1 Introduction

If the data we have is a series of numerical values that pertain to a sequence of points of time, then it is a *time series*. Daily high temperatures, the quarterly sales of a business, the annual returns of a stock and the weekly Dow Jones industrial average are all time series data.

The most common goal in time series analysis is forecasting. Given past data on sales, stock price or an index we would like to predict the value for one or more future periods. We shall see how spreadsheets can be used in forecasting.

12.2 Forecasting

In essence, forecasting amounts to finding a pattern in the past data and extending it into the future. In the case of sales data, the pattern could be a combination of trend, seasonal or cyclic variation and random variation shown below. The random variation is something that has no pattern and therefore cannot be predicted. We shall see how the other two are identified and used in forecasting.

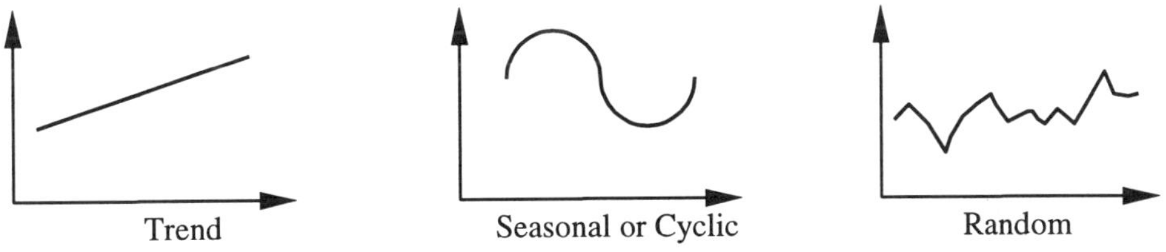

Figure 12.1.1 Common Patterns in Time Series Data

12.2.1 Trend Analysis

We have already seen trend in the regression chapter. If we fit a regression line to the sales data, then the regression line would signify the trend in sales. Trend could be positive or negative depending upon the slope of the regression line. The predicted value along the regression line extended into the future is the *trend forecast*.

On the spreadsheet, trend forecast can be made in two different ways, namely

1. using the AutoFill button
2. using the TREND function.

Figure 12.2.1 shows the sheet named "Trend" in Forecasting workbook. The data from Example 12-1 of Aczel has been entered in the range A3:B11. To find the trend forecast for year 1996 in cell B12,

- Select the range B4:B11.
- Drag the AutoFill button from B11 into B12.

Excel carries out a quick regression and makes a forecast based on the regression line. (The forecast value of 1679.61 is more accurate than the value in the textbook.) Some points to note about this method are:

1. All the selected cells must contain numerical values and not formulas.
2. The cell where the forecast appears must be directly below (or to the right, if the AutoFill button is being dragged to the right) of the selected cells. Later, it may be moved to a desired location.
3. When data is changed, the forecast will NOT update.

The other method uses the TREND function. The general format for the TREND function is

=TREND(*Y-values, X-values, New X-values, ZeroIntercept*)

The *New X-values* are the values for which *Y* will be forecast. If the New X values are a column of *n* numbers then a column of *n* forecast values will be returned.

If ZeroIntercept is set as TRUE or is omitted, the intercept will not be forced to equal zero. If it is set to FALSE, the intercept would be forced to equal zero.

An important advantage of using TREND function is that the results will update if the data is changed. We will therefore use it always.

	A	B	C	D	E
1	**Trend Analysis**				
3	**Year**	**Loans($ billions)**			
4	1990	833			
5	1991	936			
6	1992	1006			
7	1993	1120			
8	1994	1212			
9	1995	1301			
10	1996	1490			
11	1997	1608			
12		1679.61	<- Using AutoFill button		
13		1679.61	=TREND(B4:B11,A4:A11,1996)		
14					

Figure 12.2.1. Making a Trend Forecast
[Trend in Forecasting]

12.2.2 Moving-Average

An *n*-period Moving Average of a time series is the average of consecutive *n*-periods' values. One may calculate moving averages in two contexts. In the first context, it is the average of the latest *n* periods' actuals and is directly used as the forecast for the next future period. Let us say the weekly sales data and the 3-week moving average are as follows:

Week	1	2	3	4	5	6
Actual Sales	231	235	239	243	247	251
3-wk Moving Avg.			235	239	243	247

The moving average calculated at the end of the third week is the average of 231, 235 and 239 which is 235. This value is used as the forecast for week 4. When week 4 actual becomes known, the average of 2nd, 3rd and 4th week actuals are averaged to get the forecast for the 5th week and so on. If the data contains only random variations, then this simple method of forecasting would work satisfactorily. But if there is trend, as indeed there is in the values above, there will be considerable error in the forecast. The error is defined as

$$\text{Error} = \text{Actual} - \text{Forecast}$$

The error can be positive or negative. For the example above, the error is a constant +8. A trend correction is necessary to take care of this error. The above example is too simple with the data having no random variation at all, and a correction of adding 8 to the moving average would reduce the error to zero throughout. Elaborate trend correction methods exist for the more complicated usual cases, but their details are not given here.

The second context is where the moving average is calculated for a period by taking equal number of observations before and after that period. If the number of periods before and after are chosen suitably the averaging would eliminate the effect of season or cyclic, and random components because there would be equal ups and downs and they would cancel each other. The moving average then represents what the value should have been in the absence of all those distorting components.

The sheet named "Moving Avg." in Forecasting workbook is shown in Figure 12.2.2. It has the moving average calculation from Table 12-2 of Aczel. The formula in cell D5 is =AVERAGE(B4:F4) and it has been copied to the right till cell M5. (Beyond cell M5 there is not enough data.)

	A	B	C	D	E	F	G	H	I	J	K	L	M	N	O
1	**Moving Average**														
3	**t**	1	2	3	4	5	6	7	8	9	10	11	12	13	14
4	Z_t	15	12	11	18	21	16	14	17	20	18	21	16	14	19
5	**MA**			15.4	15.6	16.0	17.2	17.6	17.0	18.0	18.4	17.8	17.6		

Figure 12.2.2 Moving Average
[Moving Avg. in Forecasting]

The moving average in for period 3 is 15.4 and the actual is 11. This means that cyclic effects and random variations have brought it down to 11 from 15.4.

An n-period *weighted moving average* is one where different weights are assigned to different observations. As an example, in the 3-week moving average above, if the weight for one period old data is 0.5, two period old data is 0.3 and three period old data is 0.2, then the moving average at the end of the third week would be:

$$0.5 * 239 + 0.3 * 235 + 0.2 * 231 = 236.2$$

This weighted average of 236.2 would be used as the forecast for the fourth week. The use of weighted moving average can reduce the forecast errors, compared to simple moving average. The exact weights are determined by a trial-and-error process so as to minimize errors.

12.2.3 Exponential smoothing

Continuing with the Weighted Moving Average example above, let us say that at the end of the fourth week, we are about to make another weighted average forecast for the fifth week. But before we calculate the average we stop and note that the forecast made for the fourth week (at the end of the third week) is 236.2 and this value, being the weighted average of the past data, contains necessary information of the past. And therefore it can be used as a proxy for past data. We can save ourselves some effort by simply taking a weighted average of the fourth week actual (243) and the fourth week forecast (236.2). The weight given to the actual is denoted by w and therefore the weight for the forecast is $(1 - w)$. If $w = 0.2$, then the result would be $0.2 * 243 + 0.8 * 236.2 = 237.56$. This method of forecasting is called *exponential smoothing*. It is exponential because the weights that are assigned, in effect, to the past data form an exponential series. It is called smoothing because the method removes the effect of random and cyclic variations.

Figure 12.2.3 shows the sheet named "Exp. Smoothing". In this template, when data is entered in the shaded area the forecast values appear in the Z_t-hat column. The three columns marked |Error|, %Error and Error^2 calculate the absolute, percent and squared errors which can help the user to track the errors. In addition, they can also be used to find the best w that minimizes any of the measures of error, namely, Mean Absolute Error (MAE), Mean Absolute Percent Error (MAPE) and the familiar Mean Squared Error (MSE). The percent error is error expressed as a percentage of the actual value. MAPE is therefore better than MAE because of the comparison with the actual value. MSE is preferred when the Solver is used to optimize w. In the template the Solver is already set to minimize MSE by changing w subject to the constraints $0 \le w \le 1$. If the Solver is set to minimize MAPE, one may or may not get the best solution.

Exponential Smoothing

	MAE	MAPE	MSE
w = 0.4	11.303	1.20%	216.93

t	Z_t	$Z_t$hat	|Error|	%Error	Error^2
1	925	925			
2	940	925			
3	924	931	7	0.76%	49
4	925	928.2	3.2	0.35%	10.24
5	912	926.92	14.92	1.64%	222.61
6	908	920.952	12.952	1.43%	167.75
7	910	915.771	5.7712	0.63%	33.307
8	912	913.463	1.4627	0.16%	2.1395
9	915	912.878	2.1224	0.23%	4.5044
10	924	913.727	10.273	1.11%	105.54
11	943	917.836	25.164	2.67%	633.23
12	962	927.902	34.098	3.54%	1162.7
13	960	941.541	18.459	1.92%	340.74
14	958	948.925	9.0754	0.95%	82.364
15	955	952.555	2.4453	0.26%	5.9793
16		953.533			

If an optimal w that minimizes MSE is desired:
Unprotect the sheet.
Choose **Solver** under **Tools** menu.
Click on the **Solve** button.
When Solver is done, click the **Keep Solver Solutic**
Protect the sheet again.

Figure 12.2.3 Exponential Smoothing
[Exp. Smoothing in Forecasting]

While calculating the errors, the first two observations have been ignored, because we start with 925 in cell D6 due to lack of a forecast value there. This makes cell D7 also 925 irrespective of what *w* is. Therefore, the errors in those two cases should be ignored.

12.2.4 Seasonality and Cyclical Behavior

When seasonality exists in the time series, the most common model is:

$$Z_t = T_t * SI_t * I_t$$

where Z_t is the actual value at time t, T_t is what the actual would be if there were no seasonality or random error, SI_t is the *seasonal index* expressed as a percentage and I_t is the random error (or Irregular component). In other words, we assume a *multiplicative* model where the effects of season and random error go to multiply T_t. We shall see the steps involved in identifying T_t and SI_t for a given data of Z_t. If cyclicity exists in place of seasonality, the computation process is similar. If both seasonality and cyclicity are present then the case is more complex. But we shall assume then that the cycle is a long one so that the T_t value in the above equation contains the effect of both trend and cycle.

The steps described below have been implemented in the sheet named "Trend+Season" and is shown in Figure 12.2.4 . Exercise 12-11 from Aczel's book has been solved on this sheet. This sheet is not a template and so another problem cannot be solved on it. But the reader can see the formulas on this sheet and do similar calculations on another sheet for another problem. For a case of quarterly data instead of monthly data refer to Aczel's Example 12-3.

1. Calculate Moving Average: We calculate a *n*-period moving average as in the second context discussed above. The *n* is suitably selected to remove the seasonality or cyclicity. For instance, if we have monthly data and there is seasonality, then *n* should be 12. If we have annual data and there is a 5 year cycle, then *n* should be 5.

On the spreadsheet in Figure 12.2.4, the formula `=AVERAGE(B3:B14)` has been entered into cell C9 and it has been copied down till cell C33. (Beyond cell C33, there is not enough data.)

2. Center the Moving Average: When *n* is even, which is almost always, we need to center the moving average. If we have monthly data, and are averaging 12 months' data to remove seasonal effect, then the average would fall on month number 6.5, so to speak. (6.5 is the center of 1, 2, 3, ..., 12.) This does not align with the 6th or the 7th month's actual. But if we average two consecutive moving averages, the first one corresponding to month 6.5 and the second to 7.5, this centered moving average would fall squarely on month 7.

The formula `=AVERAGE(C9:C10)` has been used in cell D9 and it has been copied down till cell D32.

3. Calculate *SI* Values: The centered moving average at period *t* is our estimate of T_t. Dividing the actual value Zt by the moving average gives as $SI_t * I_t$ which is an approximate estimate of SI_t.

The formula `=B9/D9*100` has been entered into cell E9 and has been copied down till cell E32.

4. Average the SI Values: If we have sufficient data, then we get more than one estimate of *SI* for a given month or quarter. For example, if we have monthly data SI_7 and SI_{19} would both be of the same month. They are then averaged to get a more accurate estimate of *SI* for that month.

The formula `=AVERAGE(E9,E21)` has been entered into cell F9 and it has been copied down till cell F20.

5. Normalize the SI Values: Because of the errors, the SI values may not add up to 400 for quarterly data, or 1200 for monthly data. We therefore normalize the values by proportionately increasing them or decreasing them so that their sum is either 400 for quarterly data or 1200 for monthly data.

The formula `=SUM(F9:F20)` has been entered into cell F21 and has been copied to the right into cell G21. The formula `=F9*1200/$F$21` has been entered in cell G9 and has been copied down till cell G20. The sum in cell G21 should show up as 1200.

6. Deseasonalize the Data: Knowing the *SI* for each period, we can now *deseasonalize* the actual data, Z_t by dividing it by the corresponding *SI*. We are then left with T_t*I_t which contains only the effects of trend and random errors.

Column H has the appropriate *SI* values from the range G9:G20 copied in the right sequence. The formula `=B3/H3*100` has been entered into cell I3 and it has been copied down till cell I38.

7. Make a Trend forecast: We regress the deseasonlized values against *t*. Using the results of the regression, a trend forecast is made for one or more future periods.

The formula `=TREND(I3:I38,,37)` has been entered in cell I39.

8. Seasonalize the Trend Forecast: The trend forecast is then seasonalized to bring in the effect of season. This is done by multiplying the forecast by the appropriate *SI*.

The formula I39*H39/100 has been entered in cell B39, which yields the forecast we want.

	A	B	C	D	E	F	G	H	I
1	**Forecasting with Trend and Seasonality**							*Copied*	
2	**Month**	**Avg. Pay**						**SI**	**Deseaso**
3	Jun-92	375						99.52	376.83
4	Jul-92	370						98.87	374.22
5	Aug-92	374						99.25	376.82
6	Sep-92	378						99.74	378.97
7	Oct-92	376						99.78	376.82
8	Nov-92	380	**MA**	**CMA**	**SI-raw**	**SI-avg.**	**SI**	100.48	378.20
9	Dec-92	384	378.33	378.63	101.42	102.25	102.33	102.33	375.26
10	Jan-93	380	378.92	379.38	100.16	100.87	100.95	100.95	376.43
11	Feb-93	378	379.83	380.29	99.40	99.76	99.84	99.84	378.62
12	Mar-93	380	380.75	381.13	99.70	99.32	99.39	99.39	382.31
13	Apr-93	382	381.50	382.08	99.98	100.02	100.09	100.09	381.64
14	May-93	383	382.67	383.17	99.96	99.68	99.76	99.76	383.92
15	Jun-93	382	383.67	384.46	99.36	99.44	99.52	99.52	383.86
16	Jul-93	381	385.25	386.00	98.70	98.79	98.87	98.87	385.35
17	Aug-93	385	386.75	387.38	99.39	99.17	99.25	99.25	387.91
18	Sep-93	387	388.00	388.38	99.65	99.67	99.74	99.74	387.99
19	Oct-93	390	388.75	389.25	100.19	99.70	99.78	99.78	390.85
20	Nov-93	392	389.75	390.13	100.48	100.40	100.48	100.48	390.14
21	Dec-93	403	390.50	390.96	103.08	***1199***	***1200***	102.33	393.82
22	Jan-94	398	391.42	391.83	101.57			100.95	394.26
23	Feb-94	393	392.25	392.54	100.12			99.84	393.65
24	Mar-94	389	392.83	393.21	98.93			99.39	391.37
25	Apr-94	394	393.58	393.79	100.05			100.09	393.63
26	May-94	392	394.00	394.33	99.41			99.76	392.94
27	Jun-94	393	394.67	394.92	99.51			99.52	394.91
28	Jul-94	391	395.17	395.42	98.88			98.87	395.46
29	Aug-94	392	395.67	396.13	98.96			99.25	394.96
30	Sep-94	396	396.58	397.25	99.69			99.74	397.02
31	Oct-94	395	397.92	398.13	99.22			99.78	395.86
32	Nov-94	400	398.33	398.75	100.31			100.48	398.11
33	Dec-94	409	399.17					102.33	399.69
34	Jan-95	404						100.95	400.21
35	Feb-95	404						99.84	404.67

Figure 12.2.4. Forecasting with Trend and Seasonality
[Trend+Season in Forecasting]

12.3 Index Numbers

	A	B	C	D
1	**Price Index**			
3	BaseYear	1982	121	Base
4		**Year**	**Price**	**Index**
5		1982	121	100.0
6		1983	121	100.0
7		1984	133	109.9
8		1985	146	120.7
9		1986	162	133.9
10		1987	164	135.5
11		1988	172	142.1
12		1989	187	154.5
13		1990	197	162.8
14		1991	224	185.1

Figure 12.3.1 Index Numbers
[Index in Forecasting]

Index numbers make for quick comparison of current periods data to a base period's data. The index is calculated using the formula:

Index for period t = Period t data/Base period data * 100 %

Figure 12.3.1 shows the sheet named "Index" in Forecasting workbook. When the data is entered in the shaded area and the base period is specified, the index values are automatically calculated and displayed.

12.4 Exercises

1. Do exercise 12-2 in Aczel's book.
2. Do exercise 12-12 in Aczel's book.
3. Do exercise 12-18 in Aczel's book. In addition find the w that minimizes MSE. What is the forecast for the next period with this w?
4. Do exercise 12-27 in Aczel's book.

13. Quality Control and Improvement

13.1 Introduction

Statistics finds numerous applications in business, but its most frequent application is clearly in the area of quality control. Often, the subject itself is called Statistical Quality Control. In this chapter, we shall see a few statistical techniques routinely used in quality control.

13.2 Pareto Diagrams

The *Pareto principle* is that when there is a large list of items to be considered, just a few items would deserve most of our attention. Accordingly, when there is a large list of quality problems, just a few of them would account for about 80% of the trouble. Attacking those few items is usually an efficient strategy. A *Pareto diagram* is histogram of items, sorted in descending order of frequency, which displays both the relative and cumulative frequencies. The diagram can help a quality manager decide which problems are worth attacking first.

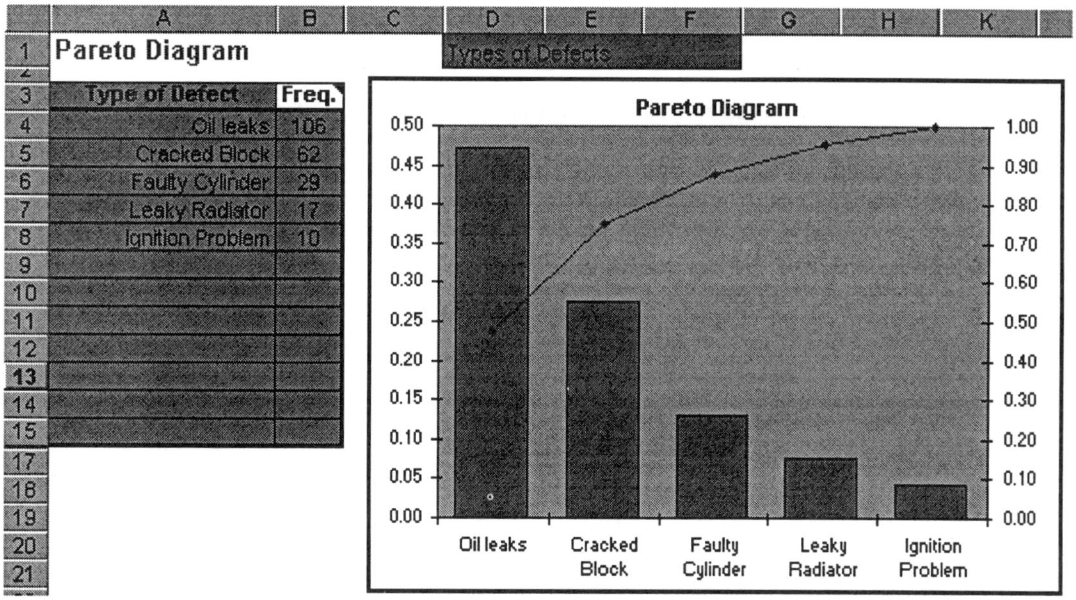

Figure 13.2.1, A Pareto Diagram
[Pareto in Quality Control]

Figure 13.2.1 shows the sheet named "Pareto" in Quality Control workbook. The currently entered data, in the range A3:B8, is from Exercise 13-10 of Aczel. While the histogram shows the relative frequencies, the line chart shows the cumulative frequencies. The cumulative frequency plot helps one to answer questions such as, "If at least 75% of the problems are to be eliminated which types of problems should be attacked?" From the cumulative frequency line, one finds the answer to be: Oil leaks and Cracked blocks. If the question is, "If Oil leaks, Cracked blocks and Faulty cylinders are taken care of, what percentage of problems would be eliminated?" From the plot, the answer is found to be: Approximately 88%.

In order to use this template, enter the data in columns A and B. If the data is not already sorted in decreasing order of the frequency,

- Unprotect the sheet.
- Select the range A4:B15.
- Select the **Sort** command under the **Tools** menu.
- Press the **OK** button.
- Protect the sheet again.

13.3 X-bar, R and s Charts

To monitor a manufacturing process, random samples of the output from the process may be taken and examined to check if the process is in control. This is nothing but a hypothesis test with the hypotheses formulated as

H_0: The process is in control
H_1: The process is out of control

And just as in the case of every other hypothesis test, there will be the chances of Type I and Type II errors. Since the process will have to be controlled continually, the hypothesis test will have to be repeated frequently. Typically, a process will be tested at the start of every shift. Thus the test needs to be conducted more than once a day, for each of the processes to be controlled. Carrying out the full set of hypothesis test computations this often is tedious. To routinize the task, *control charts* are used. A control chart contains an upper and lower limit for whatever is the test statistic Whenever a limit is exceeded, the process is declared "out of control" and corrective actions are taken.

A Type I error here will be declaring the process out of control when it actually is in control. The cost of the error comprises the lost production time and the cost of (unnecessarily) resetting of the machine. A Type II error will be not noticing that the process is out of control. The cost of this error comprises the cost of defective parts produced. The control limits on the control chart should be adjusted to reflect these costs.

Control charts are classified as variable charts and attribute charts. A variable chart is typically for controlling a process mean measured in some unit as length, weight etc. An attribute chart controls an attribute such as % defective.

We shall first see the variable control charts. These are $\overline{X}$ charts, *R* charts and *s* charts, where $\overline{X}$ stands for sample mean, *R* for sample range and *s* for sample standard deviation. These charts were developed in the 1920's when computers were not available and therefore computation of standard deviation was considered tedious. Thus the limits for $\overline{X}$ were calculated using range rather than standard deviation, and similarly for the *R* chart. The combination of $\overline{X}$ and *R* charts are common even today. With the advent of computers and spreadsheets, we might soon see the limits calculated differently, and *s* chart replacing *R* chart.

The Upper Control Limit (UCL) and the Lower Control Limit (LCL) for $\overline{X}$ are given by

$$\text{UCL} = \overline{\overline{X}} + A_2\,\overline{R}$$
$$\text{LCL} = \overline{\overline{X}} - A_2\,\overline{R}$$

where $\overline{\overline{X}}$ is the average of all sample means over a period, A_2 is a factor that depends on sample size n, and is read off of a lookup table (in Appendix C of Aczel), and $\overline{R}$ is the average of all sample ranges over a period. The center line of the chart is set at $\overline{\overline{X}}$.

The center line for the *R* chart is at $\overline{R}$ and the UCL and LCL are given by

$$\text{UCL} = D_3\,\overline{R}$$
$$\text{LCL} = D_4\,\overline{R}$$

where D_3 and D_4 depend on n, and are read off from a lookup table.

The center line for the *s* chart is at $\overline{s}$ and the UCL and LCL are given by

$$\text{UCL} = B_3\,\overline{s}$$
$$\text{LCL} = B_4\,\overline{s}$$

where $\overline{s}$ is the average of all sample standard deviations, and B_3 and B_4 depend on n, and are read off from a lookup table.

Figure 13.3.1 shows the sheet named "X-bar R s Charts" in Quality Control workbook. This template can create, $\overline{X}$, *R* and *s* charts for up to 15 samples (subgroups) and up to 10 items in each sample. The *s* chart contained in the template is not shown in the figure (in order to shorten the figure).

X-bar, R and s Charts

	1	2	3	4	5	6	7	8	9	10	11	12	13	14	15
1	121	125	129	122	119	125	121	123	128	120					
2	122	123	123	120	118	139	122	127	129	128					
3	121	121	122	121	121	150	120	123	122	120					
4															
5															
6															
7															
8															
9															
10															
x-bar	121	123	125	121	119	138	121	124	126	123					
R	1	4	7	2	3	25	2	4	7	8					
s	0.58	2	3.79	1	1.53	12.5	1	2.31	3.79	4.62					

n	3	*x*-bar-bar	124	*R*-bar	6.3	*s*-bar	3.31
		UCL	131	UCL	16.2	UCL	8.51
		LCL	118	LCL	0	LCL	0

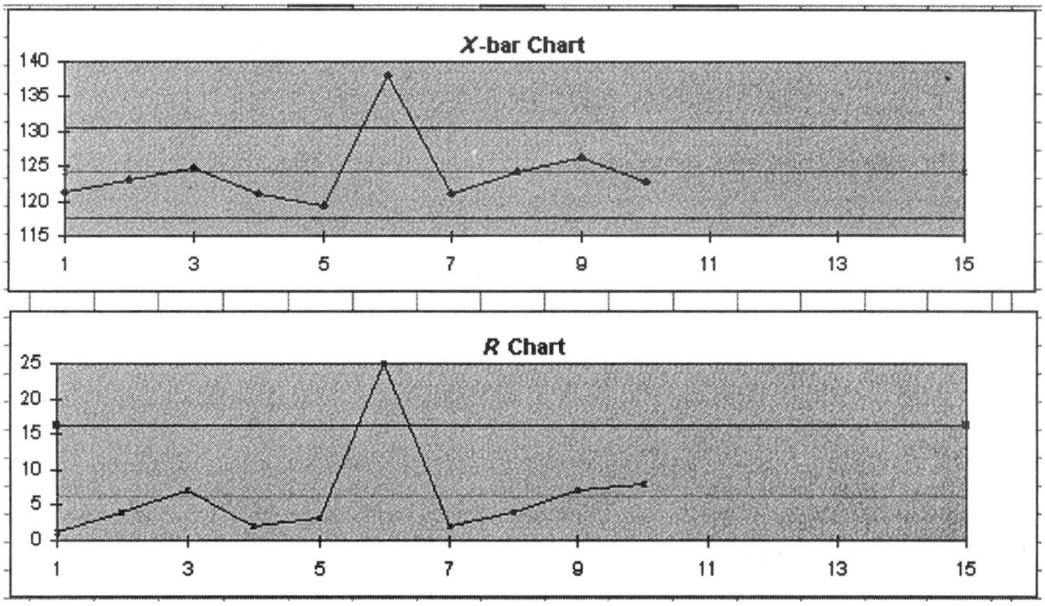

Figure 13.3.1. $\overline{X}$ and R Charts

These charts are constructed on a rolling basis, i.e., as more and more samples are taken they are added to the data in the empty columns on the right. When 15 subgroups have been collected, there is no space for the 16th. At that time, the first period data will be deleted, the remaining data will be moved one column to the left making room for the 16th period data.

The control charts can then be visually inspected for indications of process being out of control. These tests are:

1. Points exceeding the control limits. [In the case of *R* and *s* charts, it is good if the LCL is exceeded because it only means that the variations are less. It may still be worthwhile to stop and test why the extraordinarily good thing happened.]
2. Nine points in a row on one side of the center line.
3. Six points in a row steadily increasing or decreasing.
4. Fourteen points in a row alternating up and down.

5. Two out of three points in a row beyond 2σ from center.
6. Four out of five points in a row beyond 1σ from center.
7. Fifteen points in a row within 1σ from the center.
8. Eight points in a row on both sides of the center line, all beyond 1σ.

13.4 p Charts

A *p* chart is for controlling a population proportion, typically the proportion of defectives. The data needed are sample proportions collected over a period of data. The center line on the control chart is at $\bar{p}$ which is the average of all sample proportions and the UCL and LCL are given by

$$\text{UCL} = \bar{p} + 3\sqrt{\frac{\bar{p}(1-\bar{p})}{n}}$$

$$\text{LCL} = \text{Max}\ [0,\ \bar{p} - 3\sqrt{\frac{\bar{p}(1-\bar{p})}{n}}\ ]$$

The LCL cannot go below zero, which explains the formula above.

Figure 13.4.1 shows the sheet named "p Chart" which creates a *p* chart when data is entered in the shaded area.

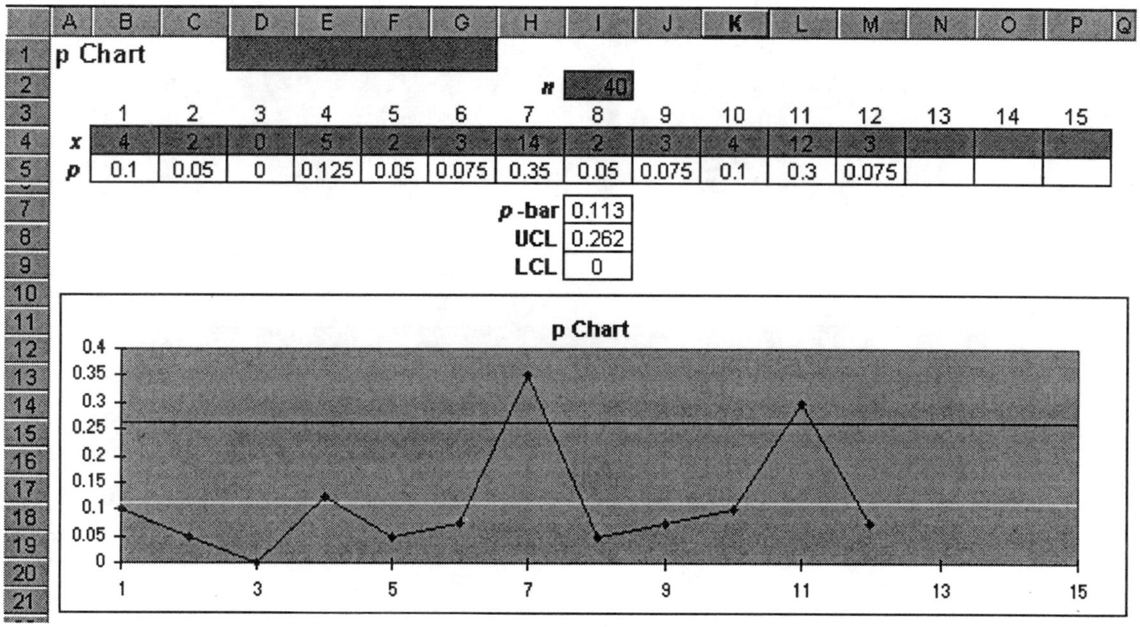

Figure 13.4.1 A *p* Chart
[p Chart in Quality Control]

13.5 c Charts

A *c* chart is useful to monitor the occurrence of mistakes that occur during a task. The number of mistakes are assumed to be Poisson distributed. The center line for the *c* chart is $\bar{c}$ which is the average number of mistakes. The UCL and LCL are given by

$$\text{UCL} = \bar{c} + 3\sqrt{\bar{c}}$$

$$\text{LCL} = \bar{c} - 3\sqrt{\bar{c}}\ .$$

Figure 13.5.1 shows the sheet named "c Chart" in Quality Control workbook. When data is entered in the shaded area of this template, a *c* chart is automatically created.

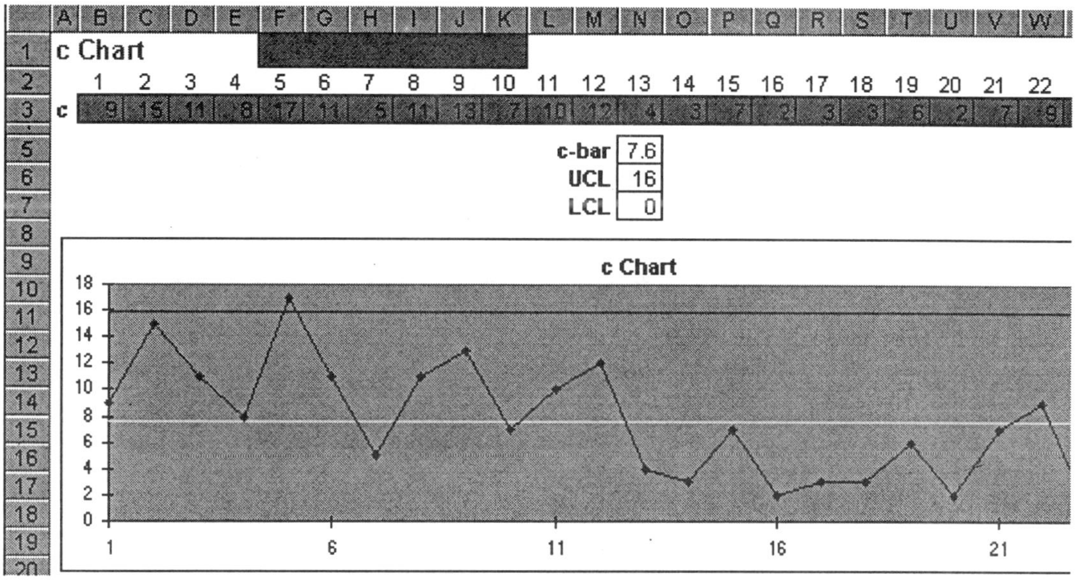

Figure 13.5.1 A c Chart
[c Chart in Quality Control]

13.6 Exercises

1. Do exercise 13-14, 21 and 24 in Aczel's book.
2. Do exercise 13-28 in Aczel's book.
3. Do exercise 13-32 in Aczel's book.

14. Nonparametric Methods

14.1 Introduction

So far, our hypothesis tests had, or were assumed to have, well defined population parameters such as mean and variance. When the data is ordinal, such as the ranks of athletes, the mean and the variance are not meaningful or not well-defined. Also in the previous chapters, the populations were normal, or assumed to be normal. When the populations do not have well-defined parameters, or when some key assumptions such as normal population fail, we resort to *nonparametric methods.*

Nonparametric methods of hypothesis testing are weaker than the parametric methods in that for the same sampling effort the nonparametric method will have more chances of Type I and Type II errors. Hence, if a parametric method is possible, it should preferred to a nonparametric method.

14.2 Sign Test

Some types of nonparametric hypothesis tests can be reduced to testing the number of + and - signs. In other words, a null hypothesis may be equivalent to "the number of + signs = the number of - signs." The test is then called a *sign test.* This test is based on binomial distribution, i.e., a binomial formula yields the *p*-value of the evidence. Figure 14.2.1 shows the sheet named "Sign" in Nonparametric Tests workbook. When data is entered in column B, the *p*-values appear in F3:H3.

A Cox and Stuart test for trend or for correlation is also a sign test. The same template can be used.

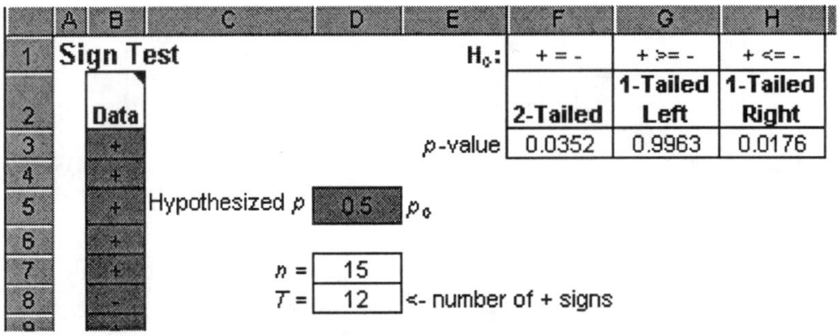

Figure 14.2.1. Sign Test
[Sign in Nonparametric Tests]

14.3 McNemar Test

Figure 14.3.1 shows the sheet named "McNemar" in Nonparametric Tests workbook.

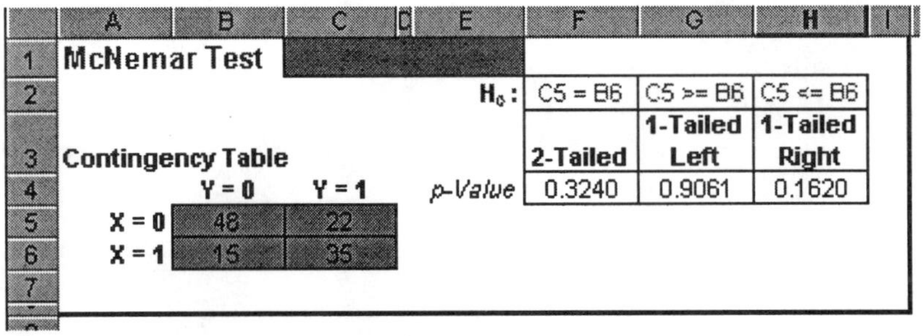

Figure 14.3.1. McNemar Test
[McNemar in Nonparametric Tests]

This template can be used to test the existence of differences in two qualitative variables X and Y. Typically, X refers to "before" and Y to "after" some treatment. $X = 0$ may refer to a subject saying "No" to a question before the treatment and $Y = 1$ may refer to "Yes" after the treatment. Thus the case of $X = 0$ & $Y = 1$ signifies the subject changed his/her mind after the treatment. This refers to cell C5 in the figure. Similarly the cell B6 refers to the number of subjects who changed from Yes to No after the treatment. The other two input cells signify the number of subjects who did not change their minds after the treatment. The hypotheses are about the number of subjects who changed from Yes to No and those who changed from No to Yes.

The p-values are calculated using the binomial distribution.

14.4 Runs Test

If we have a sequence of symbols that are, say, + or -, then a *run* is a subsequence of like symbols. For instance, the sequence

+ --++++ - ++ ---- +++ --++

has 9 runs:

+	--	++++	-	++	----	+++	--	++
1	2	3	4	5	6	7	8	9

Some hypothesis tests about randomness concern the number of runs in a sequence. The null hypothesis would imply that the number of runs should be neither too small nor too large. Thus it is usually a two tailed test. Figure 14.4.1 shows the sheet named "Runs".

The Wald-Wolfowitz test for equality of two distributions is also a Runs test. The same template can be used.

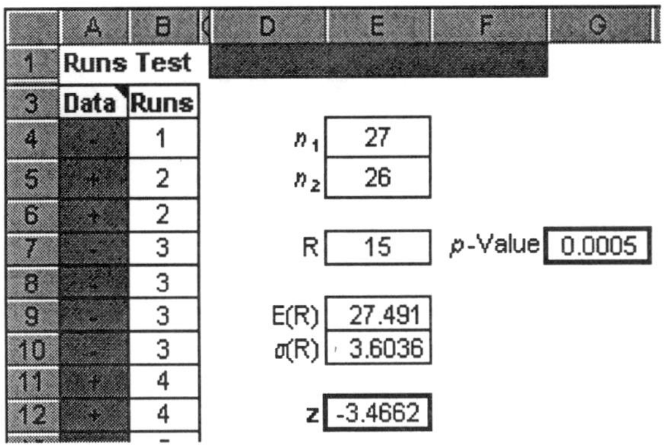

Figure 14.4.1. Runs Test
[Runs in Nonparametric Tests]

14.5 Mann-Whitney U Test

Mann-Whitney U test is the counterpart of a t-test for equality of two population means. If the assumption of normal populations fails in a t-test, then Mann-Whitney U test is the best alternative. This and a few of the subsequent tests use the ranks of data. The ranks are calculated slightly differently than the standard practice. Suppose we need to rank the numbers

132 135 135 146 154 154 154 160

in ascending order. The standard practice is to rank them as

1 2 2 4 5 5 5 8

What we need though is for tied values to have their ranks equal to the average of the tied positions. In other words, the two 135's should have rank equal to the average of 2 and 3 which is 2.5. Similarly the three 154's should have a rank of 6 each. The RANK function available in Excel only gives the standard ranks. Hence a special function has been used in all rank calculations.

Figure 14.5.1 shows the sheet named "Mann-Whitney" in Nonparametric Tests workbook.

Mann-Whitney U Test

	Group	Data	Rank
4	1	35	5
5	1	38	8
6	1	40	10
7	1	42	12
8	1	41	11
9	1	36	6
10	2	29	2
11	2	27	1
12	2	30	3

	Counts
1	6
2	6

	Rank Sums	U
1	52	5
2	26	

E[U]	18
σ(U)	

H_0:	$\mu_1 = \mu_2$	$\mu_1 <= \mu_2$	$\mu_1 >= \mu_2$
	2-Tailed	1-Tailed Left	1-Tailed Right
p-Value			

Look up U tables for p-values

Figure 14.5.1. Mann-Whitney *U* Test
[Mann-Whitney in Nonparametric Tests]

For small sample sizes, one needs to use tables in the textbook to find the *p*-values. Since the table has only the left tail of the *U* distribution, a 1-tail test on the right should be turned into a left tail one by switching the labels for population 1 and 2. For large samples where both n_1 and n_2 are greater than 10, the template provides the *p*-values. There is no need to switch to left tail. The formulas in this technique are:

$$U = n_1 n_2 + n_1(n_1 + 1)/2 - R_1.$$

$$E[U] = n_1 n_2 / 2$$

When the samples are large,

$$\sigma_U = \sqrt{\frac{n_1 n_2 (n_1 + n_2 + 1)}{12}}$$

14.6 Wilcoxon Signed Rank Test

Wilcoxon Signed-Rank test is the counterpart of paired difference t-test for difference in two population means. The paired differences are taken, the ties are discarded, and the remaining differences are ranked. A statistic denoted by *T* is computed. For a 1-tailed test on the left, *T* is the sum of the ranks of positive differences; for a 1-tailed test on the right, *T* is the sum of the ranks of negative differences; for a 2-tailed test, *T* is the minimum of the two sums. For small samples one needs to look up tables in the textbook. Figure 14.6.1 shows the sheet named "Wilcoxon".

Wilcoxon's Signed Rank Test

	X_1	X_2	Diff	|Diff|	Rank
4	56	40	16	16	9
5	48	70	-22	22	12
6	100	60	40	40	15
7	85	70	15	15	8
8	22	8	14	14	7
9	44	40	4	4	2
10	35	45	-10	10	6

Σ(+)	86
Σ(-)	34
n	15
E[T]	60
σ(T)	

H_0:	$\mu_1 = \mu_2$	$\mu_1 >= \mu_2$	$\mu_1 <= \mu_2$
	2-Tailed	1-Tailed Left	1-Tailed Right
T	34	86	34
p-Value			

Look up T tables for p-Values

Figure 14.6.1. Wilcoxon Signed-Rank Test
[Wilcoxon in Nonparametric Tests]

When the sample size exceeds 25, normal approximation can be used. The formulas for mean and standard deviations then are

$$E[T] = n(n + 1)/4$$

$$\sigma_T = \sqrt{\frac{n(n+1)(2n+1)}{24}}$$

where n is the number of observations excluding ties.

14.7 Kruskal-Wallis Test

Kruskal-Wallis test is the counterpart of single factor ANOVA. When the assumption of normal populations in ANOVA fails, this test is the best alternative. Figure 14.7.1 shows the sheet named "Kruskal-Wallis". When the sample size in every group exceeds 5, the test statistic H follows a chi-square distribution with $k - 1$ degrees of freedom where k is the number of groups. A chi-square test can then be applied. If the sample sizes are not large enough, special tables must be used to find the p-value. The formula for H is

$$H = \frac{12}{n(n+1)}\left(\sum_{j=1}^{k}\frac{R_j^2}{n_j}\right) - 3(n+1)$$

where n_j is the sample size of group j, n is the sum of all the sample sizes, and R_j is the sum of all the ranks belonging to group j.

Kruskal-Wallis Test

Group	Data	Rank		Group	R	n		
1	14	14		1	90	6	H	12.3626
1	10	10		2	56	6		
1	16	16		3	25	6	p-Value	0.0021
1	17	17						
1	15	15						
1	18	18						
2	8	8						
2	11	11						
2	7	7						
2	13	13						
2	5	5			*Total*	18		

Figure 14.7.1. Kruskal-Wallis Test
[Kruskal-Wallis in Nonparametric Tests]

14.8 Friedman Test

Friedman test is the counterpart of randomized block design test in ANOVA, when the populations are not normal. The data in each row is ranked and we shall assume the ranking has been done. The input data for the template are the ranks. The test statistic X^2 follows a chi-square distribution with $(k - 1)$ degrees of freedom where k is the number of groups. The formula for X^2 is

$$X^2 = \frac{12}{nk(k+1)}\left(\sum_{j=1}^{k} R_j^2\right) - 3n(k+1)$$

where n is the number of observations made in each of the k groups and R_j is the sum of the ranks that belong to group j.

Figure 14.8.1 shows the sheet named "Friedman". This template can take a maximum of 10 groups.

	A	B	C	D	E	F	G	H	I	J	K	L	M	N
1	Friedman Test													
2														
3	R	31	21	38	0	0	0	0	0	0	0		n	15
4		1	2	3	4	5	6	7	8	9	10	RowSum	k	3
5		1	2	3								6		
6		2	1	3								6	χ^2	9.73333
7		1	3	2								6		
8		2	1	3								6	p-Value	0.0077

Figure 14.8.1. Friedman Test
[Friedman in Nonparametric Tests]

14.9 Spearman Rank Correlation Coefficient

	A	B	C	D	E	F	G	H	I	J	K
1	Spearman Rank Correlation Coefficient										
2								H_0:	$\rho_s = 0$	$\rho_s >= 0$	$\rho_s <= 0$
3	X	Y	RankX	RankY					2-Tailed	1-Tailed Left	1-Tailed Right
4	220	151	7	6		n	10	p-Value			
5	218	150	5	5		ρ_s	0.9758		Look up the tables for p-values		
6	216	148	3	3							
7	217	149	4	4							
8	215	147	2	2		z					

Figure 14.9.1. Spearman Rank Correlation Coefficient
[Spearman in Nonparametric Tests]

For ordinal data, the correlation coefficient we saw in the regression chapter does not apply. Instead, the correlation coefficient should be applied to the *ranks*. It is then called *Spearman Rank Correlation Coefficient* and denoted by ρ_s. Also the t-test for $\rho = 0$ cannot be applied as we did before. A different one implemented in the template must be used. Figure 14.9.1 shows the sheet named "Spearman". For small samples, a table found in the textbook must be used to find p-values. For large samples ($n > 30$), the template uses normal approximation and provides the p-value.

14.10 Chi-Square Test for Goodness-of-Fit

	A	B	C	D	E	F	G	H	I	J	K	L	M
1	Chi-Square Test for Goodness of Fit												
2					Frequency data								
3	Actual	14	20	18	19	16	15						χ^2
4	Expected	15.9	17.1	17	17	17.1	15.9						1.11755
5											p-Value		0.9525
6	k	6											
7	df	5	You must enter the df here.										
8			Note:										
9			Usually, df = k-1										
10			If this is a test for normality and										
11				μ and σ are unknown then df = k-3									

Figure 14.10.1. Chi-Square Test for Goodness-of-Fit
[Chi-Sq. for Fit in Nonparametric Tests]

This test can be used when we want to test if a given data follows a desired distribution. The distribution range is divided into a convenient number of intervals, and the *observed frequency* (O_i) in each interval is calculated. The *expected frequencies* (E_i)are then calculated according to the desired distribution. The test statistic is $\Sigma\ (O_i - E_i)^2/E_i$ and it follows a chi-square distribution. The degrees of freedom depends on whether the mean and variance of the desired distribution were known. If they were known, then $df = k - 1$, else two more degrees of freedom are lost and $df = k - 3$.

Figure 14.10.1 shows the sheet named "Chi-Sq. for Fit".

14.11 Chi-Square Test for Independence

Here we test if there is any dependency among the two attributes in a cross tabulation. The test statistic is once again $\Sigma\ (O_i - E_i)^2/E_i$. It follows a chi-square distribution with $(r - 1)(c - 1)$ degrees of freedom where r is the number of rows and c is the number of columns in the cross tabulation. A special function called CHITEST is worth noting here. The formula `=CHITEST(Observed,Expected)` would return the p-value, where `Observed` and `Expected` are the ranges where the observed and expected frequencies are. Figure 14.11.1 shows the sheet named "Chi-Sq for Ind.".

	A	B	C	D	E	F	G	H	I	J	K	L	N	O	P
1	**Chi-Square Test for Independence**														
2		**Frequencies Data**													
3		1	2	3	4	5	6	7	8	9	10	Total			
4	1	21	27	40	15							103	#Rows	5	
5	2	14	18	28	8							68	#Cols	4	
6	3	11	13	21	7							52	*df*	12	
7	4	3	33	30	9							75			
8	5	30	11	34	10							85	χ^2	33.958	
9	6											0	*p*-Value	0.0007	
10	7											0			
11	8											0			
12	9											0			
13	10											0			
14	Total	79	102	153	49	0	0	0	0	0	0	383			

Figure 14.11.1. Chi-Square Test for Independence
[Chi-Sq for Ind. in Nonparametric Tests]

14.12 Exercises

1. Do exercise 14-1 in Aczel's book.
2. Do exercise 14-3 in Aczel's book.
3. Do exercise 14-5 in Aczel's book.
4. Do exercise 14-6 in Aczel's book.
5. Do exercise 14-11 in Aczel's book.
6. Do exercise 14-12 in Aczel's book.
7. Do exercise 14-15 in Aczel's book.
8. Do exercise 14-20 in Aczel's book.
9. Do exercise 14-26 in Aczel's book.
10. Do exercise 14-27 in Aczel's book.
11. Do exercise 14-40 in Aczel's book.
12. Do exercise 14-46 in Aczel's book.
13. Do exercise 14-51 in Aczel's book.
14. Do exercise 14-55 in Aczel's book.
15. Do exercise 14-61 in Aczel's book.
16. Do exercise 14-64 and 65 in Aczel's book.
17. Do exercise 14-70 in Aczel's book.

15. Decision Analysis

15.1 Introduction

Decision analysis addresses the problem of choosing the best alternative among available ones where the outcomes of each choice are uncertain and the choice has to be made before the uncertainties are resolved. Where to invest, whom to hire and which vendor to select are typical problems amenable to decision analysis.

15.2 Decision Analysis without Probabilities

A decision analysis problem is characterized by the *Payoff Table* which tabulates the outcome of each combination of decision alternatives and uncertain events. Figure 15.2.1 shows the sheet named "Without Probs" in Decision Analysis workbook. The sheet contains a payoff table with rows marked d1, d2 and d3 which are the *decision alternatives*. The columns s1-s4 are *states of nature* one of which is going to happen but we (the decision makers) do not know which. The number 80 in the first cell of the table signifies if we choose d1 and s1 happens then we receive a payoff of 80, and so on. Some payoffs are as high as 90 and some as low as 12. Thus there is risk in the situation.

Different people may decide differently in a situation like this, and indeed there is no best answer. We can say, however, according to such-and-such criterion, such-and-such decision is the best.

	A	B	C	D	E	F	G	H	I	K	L	M	N	O	P
1	**Decision Analysis without Probabilities**														
2									α						
3		**Payoff Table**							0.6	**Opportunity Loss Table**					
4		**s1**	**s2**	**s3**	**s4**	Max	Min	La Place	Hurwicz		**s1**	**s2**	**s3**	**s4**	Max
5	**d1**	80	61	50	33	80	33	56	61.2	**d1**	0	24	6	57	57
6	**d2**	12	35	52	90	90	12	47.25	58.8	**d2**	68	50	4	0	68
7	**d3**	42	85	56	40	85	40	55.75	67	**d3**	38	0	0	50	50
8															
9	**Max**	80	85	56	90	90	40	56	67					**Min**	50
10	**d***	d1	d3	d3	d2	d2	d3	d1	d3					**d***	d3

Figure 15.2.1. Decision Analysis without Probabilities
[Without Probs in Decision Analysis]

First, let us consider the optimistic approach. An optimist would look at d1 and say that if d1 is chosen, s1 would happen because that has the maximum payoff in that row. Thus, d1 is worth 80. Similarly, d2 is worth 90 and d3, 85. Selecting the best among these, an optimist will choose d2 and expect a payoff of 90. On the spreadsheet, cell F9 shows the 90 and cell F10 shows d2. The maximum in each row, and the maximum of all these maxima (in cell F9) have been calculated using the MAX function. The formula in cell F10 is

```
=INDEX($A$5:$A$7,MATCH(F9,F5:F7,0)).
```

It finds which number in range F5:F7 matches the maximum and returns d1, d2 or d3 accordingly. This formula has been copied into other cells in that row to the left and to the right.

We shall denote the best choice by d*. In this approach, the maximum along each row was picked and then the maximum among all the maxima is picked. Therefore, this decision criterion is called the *maximax criterion*, and we say "By the Maximax criterion, d* = d2."

A pessimist will look at d1 and say that if d1 is chosen s4 would happen yielding a payoff of only 33. In other words, the minimum in each row is looked at. Continuing similarly, a pessimist will at the end pick d3 seen in cell G10. In other words, "By maximin criterion, d* = d3."

Besides maximum and minimum, one may look at the average of all the payoffs in a row. The French philosopher and mathematician LaPlace put forth the principle that if we do not know details about different possibilities, then we should give them all equal probabilities. When we take the average of all payoffs we are, in effect, calculating an expected payoff with equal probabilities for each one of s1-s4.

Hence this criterion is called the *LaPlace criterion.* From the spreadsheet we see that, "By LaPlace criterion, d* = d1."

On can also take a weighted average of the maximum and minimum and use that as the criterion. This method is credited to the Nobel Prize winning economist Hurwicz. The weight given to the maximum, the optimistic expectation, is called the *coefficient of optimism* and is denoted by α. The minimum gets a weight of $(1 - \alpha)$. With an α of 0.6, the best decision turns out to be d3. Thus, "By Hurwicz criterion, d* = d3."

Yet another rational approach is to look at the *regret.* If d1 is chosen and s2 happens and we get a payoff of 61, we would regret our choice of d1 since d3 would have fetched 85. Our regret is measured by the difference 85 – 61 = 24. For each combination of s and d, a regret can similarly be calculated. These are tabulated in the Regret Table found in the range L5:O7. Now we can calculate the maximum regret in each row and then pick the alternative with the minimum among all the maximum regrets. This works out to d3. Thus, "By *minimax regret criterion,* d* = d3."

The formula used to calculate regret in cell L5 is `=$B$9-B5`, and it has been copied into the two cells below it. Similar formulas are used in the other cells of the regret table. This sheet can be used for another problem only if that problem contains three alternatives and four states of nature. Later, we shall see a template that accommodates more alternatives and states of nature.

15.3 Decision Analysis with Probabilities and Additional Information

When probabilities of s1-s4 are available, we can extend our analysis using expected values. Figure 15.3.1 shows the sheet named "With Probs & Info."

Decision Analysis

Payoff Table	s1	s2	s3	s4	EV	EV(.|I1)	EV(.|I2)	EV(.|I3)
d1	80	61	50	33	58.65	57.35	47.82	67.85
d2	12	36	52	90	42.30	42.27	60.21	28.59
d3	42	85	56	40	55.20	67.71	51.68	49.30
Prob.	0.30	0.20	0.35	0.15				
Max	80	85	56	90	58.65	67.71	60.21	67.85

Regret Table	s1	s2	s3	s4	EOL
d1	0	24	6	57	15.45
d2	68	50	4	0	31.8
d3	38	0	0	50	18.9
Prob.	0.30	0.20	0.35	0.15	

Revised Probabilities	s1	s2	s3	s4
P(.|I1)	0.107	0.500	0.313	0.080
P(.|I2)	0.096	0.096	0.448	0.360
P(.|I3)	0.589	0.074	0.301	0.037

Conditional Probabilities	s1	s2	s3	s4
P(I1|.)	0.1	0.7	0.25	0.15
P(I2|.)	0.1	0.15	0.4	0.75
P(I3|.)	0.80	0.15	0.35	0.10

Decision Strategy			
Info.	I1	I2	I3
Prob.	0.280	0.313	0.408
d*	d3	d2	d1
EV	67.71	60.21	67.85

EV with PI	EV with SI
74.10	65.43

JOINT PROBABILITY TABLE	s1	s2	s3	s4	Total
I1	0.030	0.140	0.088	0.023	0.280
I2	0.030	0.030	0.140	0.113	0.313
I3	0.240	0.030	0.123	0.015	0.408
Total	0.30	0.20	0.35	0.15	1

EVPI	EVSI	Efficiency
15.45	6.775	43.85%

Figure 15.3.1. Decision Analysis with Probabilities and Additional Information [With Probs & Info in Decision Analysis]

The probabilities of s1-s4 are in the range B9:E9. The expected value (EV) of each alternative is calculated with these probabilities in the range F5:F7. The formula in cell F5 is

=SUMPRODUCT(B5:E5,$B$9:$E:9)

and it has been copied down to two more cells. If we wish to maximize EV, our choice would then be d1 where the EV is maximum. Thus, "By Max EV criterion, d* = d1."

We can also calculate expected regret and minimize it. This is known as *minimum expected opportunity loss* (Min EOL) criterion. Calculations in the range O5:O7 reveal, "By Min EOL criterion, d* = d1." We note that the best choice by Max EV and Min EOL criteria are the same. This is not by accident; they will be the same in every problem. The reason is that the sum of the EV of an alternative and the EOL of the same alternative is constant. For instance, EV of d1 + EOL of d1 = 58.65 + 15.45 = 74.10. It is the same for d2 and d3. Thus EV and EOL are complements of each other, and maximizing one will minimize the other.

We shall next consider the possibility of getting an additional information which is *perfect*. Perfect information (PI) means all uncertainties are removed and we know exactly what is going to happen. If we come to know through a "perfect information source" that s1 is going to happen, we would choose d1 which gives the maximum payoff of 80 under s1. Similarly, we would choose d3 under s2 and s3 for payoffs of 85 and 56, and d2 under s4 for a payoff of 90. Thus if we have a perfect information source helping us out, we can jockey between the choices according to the information we get and earn better payoffs. The expected payoff with the perfect information (EV with PI) is therefore

=SUMPRODUCT(B9:E9,B16:E16)

which yields 74.10 in cell G22. We have seen this figure of 74.10 before. It was the sum of EV and EOL of any alternative. Thus we have

$$\text{EV(any d)} + \text{EOL(same d)} \equiv \text{EV with PI}$$

The symbol ≡ implies that the equation will hold in every problem.

Suppose now, whoever is the source of perfect information asks us for some payment, how much should we be willing to pay? We note that with PI we expect 74.10, and without any information we would choose d1 which has the maximum EV of 58.65. Thus the difference 74.10 − 58.65 = 15.45 is what we should be willing to pay at most. This figure is known as the *Expected Value of Perfect Information* or EVPI. We have seen the figure 15.45 also before. It was (the EOL of d1 which is also) the Min EOL. Thus we have another interesting result, namely,

$$\text{EVPI} \equiv \text{Min EOL}.$$

Thus, one way to calculate EVPI is to construct the regret table and calculate the Min EOL.

Perfect information is only an ideal, and cannot be obtained in practice. We shall next see the case of imperfect or Sample Information (SI). The use of SI is a little more involved. First we note that an information is simply a signal and what matters is how it is associated with actual events rather than what it says. For instance, let us say that there are two meteorologists who predict if it is going to rain tomorrow. If the first meteorologist is correct 70% of the time and the second is correct only 20% of the time, who is better? We realize that if we took the second meteorologist's yes as no and no as yes, he would be correct 80% of the time, and he is therefore better than the first!

	A	B	C	D	E	F	G	H	I
1	**Box and Marbles example for Bayes' Rule**								
2	**Marbles contained**				**Experiment:**				
3		**A**	**B**	**C**	One box is chosen at random.				
4	**Red**	4	52	12	From the chosen box a marble is drawn at random.				
5	**Green**	1	45	39					
6	Total	5	97	51					
7									
8	**Joint Probability Table**								
9		**A**	**B**	**C**					
10	**Red**	0.2667	0.1787	0.0784	0.5238				
11	**Green**	0.0667	0.1546	0.2549	0.4762				
12	Total	0.3333	0.3333	0.3333	1.0000				
13									
14	**Revised Probabilities**								
15		**A**	**B**	**C**					
16	**P(. | R)**	0.5091	0.3412	0.1497	1.0000				
17	**P(. | G)**	0.1400	0.3247	0.5353	1.0000				

Figure 15.3.2 Box and Marbles experiment
[Marbles in Decision Analysis]

Also, we need to know how to revise probabilities on receiving additional information, using Bayes' rule. Recall that Bayes' Rule is

$$P(A \mid B) = P(A \cap B)/P(B).$$

We shall take a quick detour to see the use of Bayes' rule. Figure 15.3.2 shows the sheet named "Marbles" in Decision Analysis workbook. It is meant for an experiment where one of the three boxes A, B and C is chosen at random and from the chosen box, a marble is drawn at random. The marbles are either red or green and the contents of the three boxes are shown on the spreadsheet. Note that box A contains predominantly red marbles and box C contains predominantly green.

In the absence of any further information, the probability of the chosen box being A is 1/3, being B is 1/3 and being C is 1/3. Suppose we are told the color of the marble drawn is red. Because A has proportionately more red and C has less, intuitively we see that the probability of the chosen box being A should be revised upward and that of C should be revised downward. The exact revised probabilities have been calculated using Bayes' rule through the use of a Joint Probability Table. The joint probability of box being A and marble being red is calculated in cell B10 using the formula `=(1/3)*(B4/B6)`. All other joint probabilities have been similarly calculated.

Using Bayes' Rule, the revised probabilities have been calculated in the range B16:D17. For instance, the formula in cell B16 is `=B10/$E$10` and it has been copied to the right. As expected, the revised probability of A has gone up from 1/3 and that of C has gone down. If the marble was green the revised probabilities in row 17 marked P(. | G) would apply. We use this method of revising probabilities in our decision analysis problem, to which we return now.

Let us assume that the sample information we get would give the signal I1, I2 or I3. (This is analogous to marbles having color I1, I2 or I3. The boxes in the analogy correspond to the states of nature. Thus there are four "boxes" and three colors for marbles.) The Conditional Probabilities table contains how the signals are associated with s1-s4. The value of 0.1 in cell B20 means that if s1 is going to happen then there would be 0.1 probability of getting the signal I1. (This corresponds to box s1 containing 0.1 proportion of I1 colored marbles.)

Using these conditional probabilities, the Joint Probability Table has been constructed. Some directly useful results from the Joint Probability Table are the marginal probabilities of I1, I2 and I3 that appear as row totals. Using the Joint Probability Table, the revised probabilities of s1-s4 have been calculated in the range B12:E14. Once the probabilities are revised, the original EV calculations in the range F5:F7 are no more valid. Hence the EV's under receipt of signals I1-I3 have been calculated using the corresponding revised probabilities in the range G5:I7.

We are now ready to analyze the use of SI. The Decision Strategy table summarizes how SI will be used. If signal I1 is received, the EV, after receiving I1, for each alternative (in the range G5:G7) reveals that d3 which has the maximum EV of 67.71 must be chosen. The probability of receiving I1 is the row total of the Joint Probability Table, namely, 0.280. Other values in the Decision Strategy table are similarly filled in. The table can be summarized as, "0.280 proportion of the times I1 will be received, d3 will be chosen and the EV is 67.71; 0.313 proportion of the time I2 will be received ..." This means that the expected payoff using the SI is =SUMPRODUCT(K20:M20,K22:M22) which yields 65.43 in cell O22. This has been labeled "EV with SI."

How much should we be willing to pay for this SI? We note that SI can increase our Max EV from 58.65 to 65.43. Thus the EV of SI, written EVSI, is the difference 65.43 – 58.65 = 6.78 which appears in cell I26. The ratio of EVSI to EVPI is called the Efficiency of SI. The efficiency works out to 43.85%.

Since an additional information cannot make our situation worse (we can always ignore an information) EVSI cannot be negative. EVSI cannot be better than the perfect information. Thus

$$0 \le EVSI \le EVPI$$

These calculations have been automated in a template comprising three sheets named "Data", "Results" and "Calc". Figure 15.3.3 shows the data sheet and Figure 15.3.4 shows the results sheet. The calculation sheet is not shown here. This template can accommodate up to 10 states of nature, 10 decision alternatives and 5 signals.

15.4 Decision Trees

Decision making situations that require *serial* decisions one after the other are best modeled as decision trees. Although every decision tree can be converted into a payoff table, the tree form is much more convenient for most calculations. Special purpose software that solve decision tree problems are available in the market. No spreadsheet template is presented here for decision trees.

	A	B	C	D	E	F	G	H	I	J	K	L
1	Data		To use this Template, enter the data in the shaded area. Results are shown on the r									
3	Payoff Table		Name1	Name2	Name3	Name4	Name5	Name6	Name7	Name8	Name9	Name10
4			s1	s2	s3	s4	s5	s6	s7	s8	s9	s10
5	Name1	d1	80	61	50	33						
6	Name2	d2	12	35	52	90						
7	Name3	d3	42	85	56	40						
8	Name4	d4										
9	Name5	d5										
10	Name6	d6										
11	Name7	d7										
12	Name8	d8										
13	Name9	d9										
14	Name10	d10										
16		Prob.	0.3	0.2	0.35	0.15						
17												
18	Conditional Probabilities											
19		I1	0.1	0.7	0.25	0.15						
20		I2	0.1	0.15	0.4	0.75						
21		I3	0.8	0.15	0.35	0.1						
22		I4										
23		I5										

Figure 15.3.3. The Data Sheet
[Data in Decision Analysis]

	A	B	C	D	E	F	G	H
1	Results				α			
2		Best Decision by			0.6			
3		Maximax	Maximin	La Place	Hurwicz	Minimax regret		
4	d*	d2	d3	d1	d3	d3		
5	Exp. Payoff	90	40	29	67			
6								
7		Best Decision Under						
8		No Info.	I1	I2	I3			
9	d*	d1	d3	d2	d1			
10	EV	58.65	67.7143	60.208	67.8528			
11	Prob.		0.28	0.3125	0.4075			
12								
13			EVPI	15.45				
14			EVSI	6.775				
15			Efficiency of SI	43.85%				

Figure 15.3.4 The Results Sheet

15.5 Exercises

1. In the box and marbles experiment, it would be wrong to calculate the revised probability P(A | Red) as "There are a total of 68 red marbles in all three boxes together and 4 of them are in A. Hence P(A | Red) is 4/68." Why?

2. Pizzas 'R Us is a national restaurant chain with close to a hundred restaurants across the United States. It is constantly in the process of finding and evaluating possible locations for new restaurants. For any potential site, Pizzas 'R Us needs to decide the size of the restaurant to build at a that site, namely, small, medium, or large; or whether to build none. For a particular prospective site, the accounting department estimates the present value (PV) of possible annual profit for each size as follows:

Estimated PV of annual profits ($ '000)				
			Demand	
		Low	Medium	High
	Small	48	32	30
Size	Medium	-64	212	78
	Large	-100	12	350
	None	0	0	0

i. What is the best decision by Maximin, Maximax, La Place, Hurwicz criterion with $\alpha = 0.3$ and *Minimax regret criteria?*

In the absence of any additional information, the prior probabilities for Low demand is estimated to be 0.42, and for Medium demand 0.36.

ii.. What is the best decision by Max EV criterion?
iii. What is the EVPI?

Since the EVPI is large, Pizzas 'R Us decides to gather additional information. It has two potential market researchers, Alice Miller and Becky Anderson, both of whom it has contracted many times in the past. The company has a database of the demand levels predicted by these researchers in the past and the actual demand levels realized. The cross tabulation of these records are as follows:

Alice Miller

			Actual	
		Low	Medium	High
	Low	9	4	2
Predicted	Medium	8	12	4
	High	4	3	13

Becky Anderson

			Actual	
		Low	Medium	High
	Low	12	2	1
Predicted	Medium	2	15	2
	High	0	3	20

iv. Alice Miller charges $12,500 for the research and Becky Anderson $48,000. Whom should the company contract for additional information? Why?

16. Multivariate Analysis

Multivariate analyses require intensive calculations such as the computation of eigen values of large matrices. It is difficult to do such computations on spreadsheet templates without using macros. The use of macros makes a template too much of a black box to students. As a result no template has been prepared for this chapter.

17. Sampling Methods

17.1 Introduction

In this chapter, we shall see special sampling methods that can increase the accuracy of estimation and/or decrease the cost of sampling.

17.2 Stratified Sampling

Often, the population consists of strata that have their own individual characteristics in terms of their mean, variance or cost of sampling. By spreading the sample appropriately over the different strata, we can increase accuracy and/or decrease cost.

Figure 17.2.1 shows the sheet named "Stratified for mu" which can be used for estimating population mean with a stratified sample. Let the notations be:

N_i for the size of the i-th stratum
n_i is thc size of the sample from i-th stratum,
N and n are the respective sums
W_i is the population proportion of the i-th stratum given by Ni/N
f_i is the sample proportion for the i-th stratum given by n_i/n
m is the number of strata
$\bar{x}_i$ is the mean of the sample from i-th stratum
s_i^2 is the variance of the sample from i-th stratum.

With these notations, the estimate of population mean $\bar{X}$ and its variance $S^2(\bar{X})$ are given by

$$\bar{X} = \Sigma\, W_i\, \bar{x}_i$$

$$S^2(\bar{X}) = \Sigma\, W_i^2\, s_i^2\, (1 - f_i)/n_i$$

These formulas have been employed in the template. The confidence interval has been calculated assuming large samples.

	A	B	C	D	E	F	G	H	I	J	K	L
1	**Stratified Sampling**											
3	Stratum	1	2	3	4	5	6	7	8	9	10	Total
4	**N**	100	100	150	50	50	50					500
5	**W**	0.2	0.2	0.3	0.1	0.1	0.1					
6	**n**	20	20	30	10	10	10					100
7	**f**	0.2	0.2	0.2	0.2	0.2	0.2					
8	**x-bar**	52.7	112.6	85.6	12.6	8.9	52.3					
9	**s²**	97650	64300	76990	18320	9037	83500					
11				**X-bar**	66.12							
12				**V(X-bar)**	532.58							
13				**S(X-bar)**	23.078							
15			**α**	**(1-α) CI for X-bar**								
16			5%	66.12	+ or -	45.231	or	20.889	to	111.35		

Figure 17.2.1. Stratified Sampling for Population Mean
[Stratified for mu in Sampling Methods]

In the case of population proportion, let p_i denote the proportion of the sample from i-th stratum. The population proportion P and its variance $S^2(P)$ are given by

$$P = \Sigma\, W_i\, p_i$$

$$S^2(P) = \Sigma\, W_i^2\, p_i(1 - p_i)/n_i$$

If the size of the strata are small, a finite population correction is necessary. With that correction, the formula is

$$S^2(P) = \sum N_i^2 (N_i - n_i) \frac{p_i(1-p_i)}{(N_i - 1)n_i}$$

These formulas have been implemented in the sheet named "Stratified for p" in Sampling Methods workbook. It is shown in Figure 17.2.2.

Stratified Sampling

Stratum	1	2	3	4	5	6	7	8	9	10	Total
N	650000	350000									1000000
W	0.65	0.35									
n	130	70									200
f	0.0002	0.0002									
x	28	18									
p-hat	0.2154	0.2571									

P-hat	0.23
V(P-hat)	0.0009
S(P-hat)	0.0297

α	(1-α) CI for P		
5%	0.23 + or - 0.0583	or	0.1717 to 0.2883

Figure 17.2.2. Stratified Sampling for Population Proportion
[Stratified for p in Sampling Methods]

17.3 Optimum Allocation

When unit sampling cost for each stratum along with the size and variance of each stratum are known it is possible to select n_i values that would maximize accuracy and minimize cost. This method of deciding the optimal n_i values is called *optimum allocation.*

Figure 17.3.1 shows the sheet named "Allocation". The formula for optimal f_i is

$$f_i = (W_i \sigma_i / \sqrt{C_i}) / \Sigma (W_i \sigma_i / \sqrt{C_i})$$

where C_i is the cost of sampling one item in i^{th} stratum. The optimal n_i values are then calculated using $n_i = n f_i$ where n is the desired total sample size.

Optimum Allocation

	1	2	3	4	5	6	7	8	9	10	Total	
W	0.4	0.5	0.1								1	
σ	1	2	3									
C	4	9	16									
f	0.3288	0.5479	0.1233									
n	329	548	123								1000	n

Figure 17.3.1. Optimum Allocation
[Allocation in Sampling Methods]

17.4 Cluster Sampling

Suppose we want to sample the students at an university. A convenient way is to randomly select a class, go to that class and sample the whole class. If a desired number of classes (without enrollment overlap) are sampled, we then have a fairly representative sample. This method is known as *cluster sampling* where each class is a cluster. Sampling all the elements in a cluster at once can save time and money.

Let the notations be

n_i = the size of i-th cluster
m = number of clusters sampled
M = the total number of clusters in the population
$\overline{n}$ = average cluster size = $\Sigma\, n_i \,/\, m$
$\overline{x}_i$ = average of i-th cluster.

With these notations, the formulas for the estimate of population mean $\overline{X}$ and its variance $S^2(\overline{X})$ are

$$\overline{X} = \Sigma\, n_i \overline{x}_i \,/\, \Sigma\, n_i$$

$$S^2(\overline{X}) = (M - m)/(Mm\overline{n}^2)\frac{\sum_1^n n_i^2(\overline{x}_i - \overline{X})^2}{m-1}$$

These formulas have been implemented in the sheet named "Cluster for mu" shown in Figure 17.4.1.

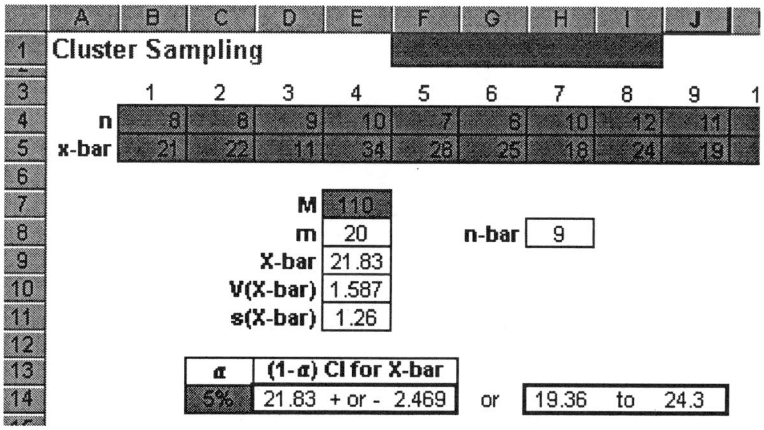

Figure 17.4.1. Cluster Sampling for Population Mean
[Cluster for mu in Sampling Methods]

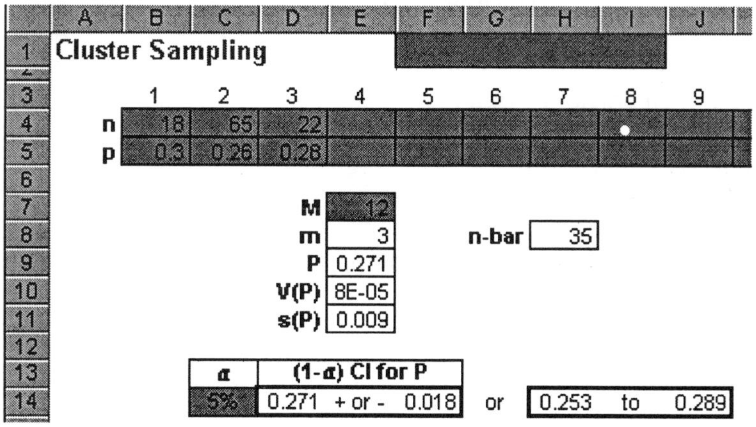

Figure 17.4.2. Cluster Sampling for Population Proportion
[Cluster for P in Sampling Methods]

The case of cluster sampling for population proportion is quite similar. The formulas for the estimate P and its variance are

$$P = \Sigma\, n_i \overline{x}_i \,/\, \Sigma\, n_i$$

$$S^2(p) = (\mathrm{M} - \mathrm{m})/(\mathrm{Mm}\bar{n}^2)\frac{\sum_1^n n_i^2(p_i - p)^2}{m-1}.$$

These formulas have been implemented in the sheet named "Cluster for P" which is shown in Figure 17.4.2.

17.5 Systematic Sampling

In systematic sampling every k^{th} item in the population is sampled, where k is selected suitably. The advantage is that the sample is spread over the population evenly, which might go to increase the accuracy of the estimate.

In this method, the sample mean is the estimate of the population mean. The variance of the estimate is given by

$$S^2(\bar{X}) = [(N - n)/Nn]\, s^2$$

This formula applies when the population has not been ordered in any particular order. This is what has been implemented in the sheet named "Systematic", shown in Figure 17.5.1. When the population has been ordered, say, in ascending order, more complicated formulas are applicable. Such formulas are not presented here.

Systematic Sampling

N	n	x-bar	s^2
2100	100	0.5	0.36

X-bar	0.5	
V(X-bar)	0.00343	When the population is in no particular order.
S(X-bar)	0.05855	

α	(1-α) CI for X-bar				
5%	0.5	+ or -	0.11476	or	0.38524 to 0.61476

Figure 17.5.1. Systematic Sampling
[Systematic in Sampling Methods]

17.6 Exercises

1. Do exercise 17-2 in Aczel's book.
2. Do exercise 17-3 in Aczel's book.
3. Do exercise 17-4 in Aczel's book.
4. Do exercise 17-6 in Aczel's book.
5. Do exercise 17-8 in Aczel's book.
6. Do exercise 17-14 in Aczel's book. Assume that the population is in no particular order.

INDEX

IRWIN/McGRAW-HILL

Statsheets with 3.5" Excel Templates and Data Files for use with Complete Business Statistics, 4/e, by Amir D. Aczel.

Please use this postage-paid form to report any errors that you find in this material. Be as complete as possible noting specifically which changes should be made. We will address them in subsequent printings and future editions. Thank You.

Attention: Scott Isenberg

Name ________________________________ School __________________________

Office Phone ___________________________

Please fold and seal so that our address is visible.

NO POSTAGE NECESSARY IF MAILED IN THE UNITED STATES

BUSINESS REPLY MAIL

FIRST-CLASS MAIL PERMIT NO. 204 OAKBROOK, IL

POSTAGE WILL BE PAID BY ADDRESSEE

ATTENTION: Scott Isenberg

THE McGRAW-HILL COMPANIES
RICHARD D. IRWIN
1333 BURR RIDGE PKY.
BURR RIDGE, IL 60521-0085

(fold)

(fold)